# Studies in Ecological Economics

Volume 9

**Series editor**
R. Kerry Turner, University of East Anglia School of Environmental Sciences, Norwich, UK

The series "Ecological Economics" aims to present monographs and edited volumes that offer fresh perspectives and overviews on a range of topics in Ecological Economics. Special attention will be given to integrative research methods and techniques and to pluralistic approaches. Fundamental and critical discussions are invited of: theoretical assumptions, ethical starting points, behavioural models, (co) evolutionary change, integration concepts, the ecosystem approach and ecosystem services environmental macroeconomics, industrial ecology, spatial dimensions, thermodynamics and production functions, policy goals and instruments, international policy dimensions, alternative welfare measures, valuation and benefits transfer. Particular emphasis will be placed on the interaction between valuation, modelling and evaluation in a multidisciplinary setting; the link between ecology, biodiversity, ecosystem services, economics politics and environmental management; the incorporation of physical flows in economic models; and the interface between development, poverty, technology and sustainability. In addition, applied and policy oriented research is welcomed, addressing specific resources, substances, materials, regions, sectors, countries or environmental problems. International comparative studies are also encouraged. The ultimate aim of this series is to present a rigorous but broad perspective on contemporary and future environmental policy questions.

More information about this series at http://www.springer.com/series/6526

R. Kerry Turner • Marije Schaafsma
Editors

# Coastal Zones Ecosystem Services

## From Science to Values and Decision Making

Springer

*Editors*
R. Kerry Turner
School of Environmental Sciences
University of East Anglia
Norwich, UK

Marije Schaafsma
Geography and Environment & Centre
for Biological Sciences
University of Southampton
Southampton, UK

ISSN 1389-6954
Studies in Ecological Economics
ISBN 978-3-319-17213-2 ISBN 978-3-319-17214-9 (eBook)
DOI 10.1007/978-3-319-17214-9

Library of Congress Control Number: 2015936343

Springer Cham Heidelberg New York Dordrecht London


Printed on acid-free paper

Springer International Publishing AG Switzerland is part of Springer Science+Business Media (www.springer.com)

*This book is dedicated to the memory of Laurence Mee. Laurence was an excellent marine scientist but also a very active contributor to the coastal and marine policy process. His boundless energy and enthusiasm was applied across a whole range of issues, with policy relevance being the key feature. He was also a very effective communicator and used this skill to good purpose around the globe in numerous academic and governance circles. On a personal level, his warm-hearted nature touched everyone he came into contact with. Laurence's passing will leave a big gap in the marine science and policy field and among his many academic colleagues who were also his friends. His memory will live long with all of us who were privileged to know and work with him over the years. Laurence was a significant contributor to the NEAFO and will be greatly missed by all of us.*

# Acknowledgments

This book was possible only through the generous support of several individuals, projects, and institutions.

The work for this book has been conducted under the UK National Ecosystem Assessment Follow-On (UK NEA FO) Work Package 4 and the Valuing Nature Network (VNN) project on Coastal Management. The UK NEA FO was funded by the Department for Environment, Food and Rural Affairs (Defra), The Welsh Government, and three research councils: the Natural Environment Research Council (NERC), the Economic & Social Research Council (ESRC), and the Arts & Humanities Research Council (AHRC). The VNN was sponsored by the Natural Environment Research Council (NERC), with additional funding from the Economic & Social Research Council (ESRC), the Department for Environment, Food and Rural Affairs, the Ecosystem Markets Task Force, and the Cambridge Programme for Sustainability Leadership. The project described in Chap. 8 was funded by the Marine Management Organisation (MMO) and Marine Scotland and managed by the Marine Environmental Data & Information Network (MEDIN). Chapter 12 was made possible through funding received from the European Community's Seventh Framework Programme (FP7/2007-2013) project "Knowledge-based Sustainable Management for Europe's Seas" (KNOWSEAS) under grant agreement n° 226675 and from the EU under the InterReg IVa-2seas programme through the project "Mnemiopsis leidy: Ecology Modelling and Observation" (MEMO).

We wish to thank colleagues within the VNN, in particular within the Coastal Management project, for providing useful comments and discussion throughout the duration of the project. A special word of thanks goes out to Steve Albon (James Hutton Institute), Sam Anson (Marine Scotland Science), Peter Barham, Peter Burbridge, Mike Cowling (Crown Estate), Deanna Donovan (JNCC), Paul Ekins (UCL), Ingvild Harkes (Edinburg Napier University), Louise Heaps (WWF), Mike Heath (University of Strathclyde), Jason Holt (National Oceanographic Centre), Dickon Howell (Marine Management Organisation), Jonathan Hughes (Scottish Wildlife Trust), Mark Huxham (Edinburg Napier University), Janet Khan-Marnie (SEPA), Mansi Konar (Defra), Aisling Lannin (Marine Management Organisation),

Marion Potschin (University of Nottingham), Dave Raffaelli (University of York), Sue Rees (Natural England), Eva Roth, James Spurgeon (Sustain Value), Selina Stead (University of Newcastle), Mavra Stithou (Marine Management Organisation), Beth Stoker (JNCC), Jamie Tratalos (University of Nottingham), David Vaughan (JNCC), and Anne Walls (BP).

Norwich, UK
Southampton, UK

R. Kerry Turner
Marije Schaafsma

# Contents

## Part III Case Studies

# Chapter 1
# Introduction

**R.K. Turner**

## 1.1 Adaptive Management in Response to Environmental Change

This volume brings together interdisciplinary analysis in order to apply the 'ecosystem services' concept and framework to coastal environments. The core objective is to outline how a more 'adaptive management' strategy (see Fig. 1.1) could be facilitated on the basis of ecosystem services thinking. The figure outlines the main steps and requirements in the management process and sets them within the European marine policy regime.

Coastal zones contain a rich diversity of ecosystems and habitats which supply a flow of valuable ecosystem services, e.g. from sea defence to carbon storage, recreation and amenity services, of great benefit in terms of welfare and well-being to society. But coasts are also subject to constant change and a sustainable service flow is dependent on the maintenance of a sufficient (quantity and quality) stock of natural capital. This stock includes ecosystems and their interrelationships and links to the abiotic environment. The stock is being put under increasing pressure and strain from a range of human developments including climate change, and the resilience properties of coastal systems need to be better understood.

The global economy driven by international finance and trade flows now impacts on almost all the world's ecosystems, not least coastal and marine, through a range of pressures and drivers. Coastal areas in particular are exposed to environmental change pressures because of their spatial location and attributes. They have therefore seen disproportionate increases in human population and economic and trade activities with consequent natural habitat destruction and coastal hard engineering

R.K. Turner (✉)
School of Environmental Sciences, University of East Anglia, Norwich, UK
e-mail: r.k.turner@uea.ac.uk

 
R.K. Turner, M. Schaafsma (eds.), *Coastal Zones Ecosystem Services*,
Studies in Ecological Economics 9, DOI 10.1007/978-3-319-17214-9_1

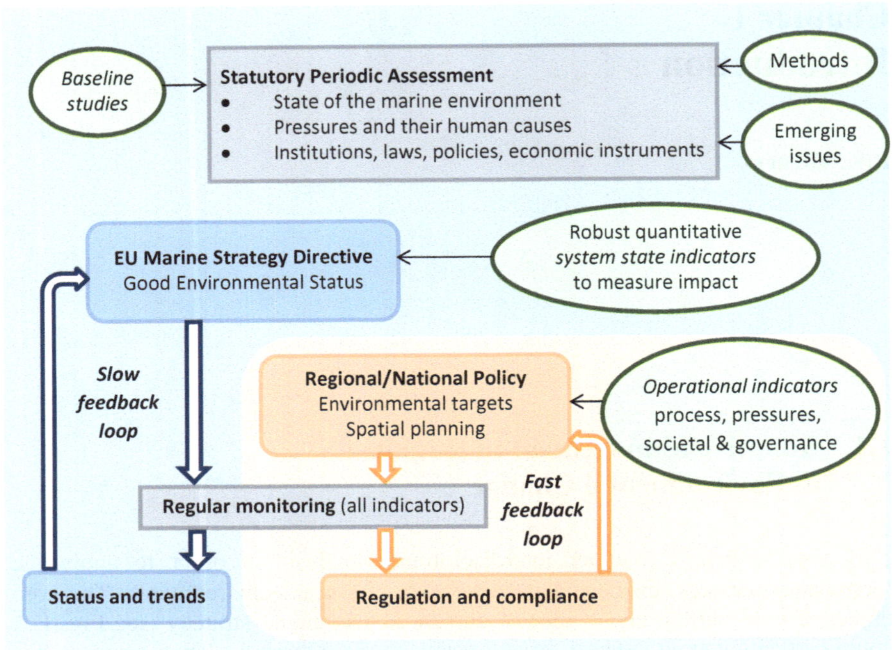


**Fig. 1.1** Scheme for adaptive management used in current EU marine policy (particularly the MSFD) (Adapted from Mee (2005))

in response to actual and perceived increased flood and erosion risk. Coastal areas and shelf seas have been the repositories for both flow pollution leading to eutrophication and stock pollution through heavy metal contamination. The sources of this pollution have been in adjacent catchments, in estuarine locations or from direct discharges into the seas. Marine litter is now a growing problem. Global trade has served to intensify shipping activities with pollution impacts and the introduction of exotic invasive species into new areas through a combination of technological changes in hull design, ballast water discharges and probably climate change effects (see Chap. 12). Offshore aggregates mining, increased navigation dredging, wind farm developments and aquaculture have also lead to habitat and species loss.

A number of definitions of the coastal zone have been proposed and the one adopted here is in line with IGBP LOICZ (Crossland et al. 2005). The coastal zone is defined as a long narrow feature consisting of mainland, islands and adjacent shelf seas, denoting the zone of transition between land and the marine domain. Coastal zones occupy around 20 % of the earth's surface but host more than 45 % of the global population and 75 % of the world's largest urban agglomerates. From a management perspective, coasts are affected by environmental changes across a range of temporal and spatial scales including the continuum from river catchment to coastal ocean. In practical terms, the definitions of the coastal zone need to vary according to the type of problem or set of issues being addressed, the prevailing governance regime and the objectives of the management process (Elliott and Whitfield 2011). Despite the focus on coastal systems due note will also be given to the interrelationships with terrestrial and deep ocean systems (Mee 2012; Barbier et al. 2014).

The adaptive management strategy championed in this book builds on some foundational principles – the 3Ps – , pluralism, pragmatism and precaution. These guiding principles are used to construct a decision support system (DSS) to help deliver a more flexible and 'learning by doing' approach to the stewarding of our precious coastal environments (see Chap. 2).

Pluralism is critical to the building of a DSS that encompasses ecosystem services because it requires collaboration across scientific disciplines. In addition, contemporary society now contains a growing diversity of social and cultural values, ethics and norms. These factors make the value of nature a multidimensional concept that is context dependent (Turner 1999). Nature's value therefore includes monetary value but also more qualitative measures. The complete 'commodification' of nature is an ever present danger to be avoided according to critics of monetary valuation (Sagoff 1988; Norgaard 2010; Baveye et al. 2013). The position adopted in this book is that many (but not all) ecosystem services can be meaningfully expressed in monetary terms and that this type of calculus has 'political' purchase which can be used to further conservation efforts in the real world (see Chaps. 4 and 6).

Pragmatism is necessary in order to raise awareness of the ecosystem services concept within governance circles (particularly in finance ministries). The underlying aim is to manage ecosystems in a way that maintains or enhances their resilience and the valuable flow of services they provide rather than to maximise biodiversity conservation per se as a moral imperative.

A precautionary approach to decision making is also recommended because of the scientific uncertainty that shrouds how the overall system and some ecosystems may be adversely affected by human development. The ever present danger of threshold effects or tipping points leading to degradation or even collapse of ecosystem functioning must be constantly borne in mind. The use of sustainability regulations and standards to protect the overall 'system' has a continuing role to play as economic development and markets continue to expand. Nevertheless, we cannot wait for more complete information in many contexts as this may result in services being further degraded or lost. Decisions will therefore need to be taken within a risk-based framework using adaptive management principles which emphasise flexibility (see Chap. 2).

Management challenges in these dynamic land and sea areas include the potentially conflicting demands on their use for different human wants and needs, together with the requirement for habitat and species conservation (Halpern et al. 2013). This cocktail often leads to highly contested management options and decisions. In the past decision makers considering a change in coastal policy at the strategic level (e.g., a switch from 'hold the line' hard engineering sea defence to a more flexible mixed soft and hard system with realigned coastal frontages, see Chap. 11) have not always fully followed through the consequences for policy delivery at the local level. Coastal management requires a tricky balance between strategic requirements and locally favoured schemes. Because of the way coastal processes often work, small scale schemes can have much wider and sometimes negative consequences. One community's erosion control scheme, for example, can mean a loss of beach frontage further down the coast. The management process will increasingly have to utilise centrally directed information and options, as well as containing the means to effectively engage with local stakeholders and their social networks to receive feedback information and alternative option suggestions. Management will therefore need to be both sensitive and responsive, and ensure that information flows between central decision makers and affected communities are as transparent as possible (see Chap. 9).

An effective DSS should be composed of a number of interrelated stages: problem/issue scoping; indicators selection (see Chap. 5), scenario analysis (see Chap. 7), modelling (see Chap. 3) and socio-economic appraisal (including monetary and non-monetary valuation, see Chaps. 4 and 6). A novel approach (the "Balance Sheets Approach") to formatting, interrogating and presenting data and findings from appraisal processes will be highlighted in order to provide as robust an evidence base as is feasible (see Chaps. 2 and 4). The aim is to enable government to better anticipate regional and local environmental implications of a change in policy and to identify where necessary practicable compensation for 'losers'. The Balance Sheets Approach is both a process and a tool, which also seeks to open up flows of information horizontally and vertically, and to make progress towards greater stakeholder influence and more co-produced policy (Turner and Welters 2014). But any DSS or changes to the system will be heavily conditioned by the prevailing governance regime.

## 1.2 Coastal and Marine Governance

Marine governance is complex and fragmented and faces formidable challenges with much of the marine domain characterised as global commons subject to open access and possible unsustainable exploitation, and protected only by international conventions. Governance in the European seas, for example, aims to both implement the Ecosystem Approach in marine management and in parallel support economic growth in the blue economy. Meeting these twin objectives will require the integration of a raft of existing EU Directives and policies at the European and national scales. The ecosystem and nature conservation objectives are encapsulated in the EU Marine Strategy Framework Directive (MSFD), the Habitats and Birds Directive (HBD) and the Water Framework Directive (WFD); while stimulation of the blue economy is encouraged through the Marine and Maritime Agenda for growth and jobs, the "Limassol Declaration" 2012.

Many countries have constructed an unnecessarily complicated marine legislation and administrative regime (Boyes and Elliott 2015). All countries have to respond to a suite of international, regional and national policies, laws and agreements covering fisheries, energy, shipping and trade and nature conservation which are enabled through multiple organisations and administrative bodies. Progress so far in terms of the necessary policy integration and coordination has been slow. There remains a lack of coordination between relevant European Commission DGs, between the Commission and international organisations, Regional Sea Convention Commissions (OSPAR, for the North East Atlantic area; HELCOM, for the Baltic; UNEP-MAP for the Mediterranean area; and the Bucharest Convention for the Black Sea) and between national Member states' policies. At the national level, within territorial waters or a country's Exclusive Economic Zone (EEZ), for example, marine spatial planning, tourism and energy exploitation are typically regulated by different government departments which then face communication and coordination problems exacerbated by the multiplicity of pieces of legislation and policy.

In the EU each member state had to define "Good Ecological Status" (GES) for its EEZ by July 2012 and to set out a programme of measures to achieve or maintain GES by 2015 (with a final deadline of 2026). Despite the obvious need for regional cooperation and coordination on trans-boundary issues, the MSFD does not provide any specific legal framework or means to ensure such activity. It also fails to provide guidance on how to engage stakeholders in the decision making and implementation process. Raakjaer et al. (2014), in line with previous UNEP thinking, have advocated a new nested governance system to address the coordination, cooperation and stakeholder participation requirements. In an ideal nested arrangement, institutions, policies, laws and sectors are encompassed within a tiered, internally consistent and mutually re-enforcing DSS and decision making process. Stakeholder influence should increase in the nested system as the process shifts from decisions about principles to implementation practice. This approach to governance has much to recommend it but the real world situation and politics in the marine context contains a number of formidable obstacles to this type of radical reform.

Raakjaer et al. (2014) suggest a polycentric model of governance which contains multiple centres of decision making which are formally independent of each other, but function coherently with consistent and predictable patterns of behaviour and feedback loops. The polycentric model requires the state and other institutions to scale down to the ecosystem level and to embrace a network orientated form of governing with vertical and horizontal linkages capable of linking together the range of stakeholder social networks and the centre. So the next question becomes how do we get to this form of governance from where we are now?

In the UK, the Marine and Coastal Access Act (2009) and the creation of the Marine Management Organisation (MMO) offered an opportunity to harmonise management. Boyes and Elliott (2015) however maintain that many overlapping responsibilities still exist, with the MMO acting as the regulator for most but not all of the marine environment and related economic activities. Van Takenhove et al. (2014) more generally have explored four possible governance models which could address to a greater or lesser extent the marine regional cooperation and coordination problem.

The simplest models are based on so-called "Advisory Alliance" (AA) and "Cross-Border Platforms" (CBP). The former envisages something similar to existing Regional Advisory Councils (RACs) for fisheries under the EU Common Fisheries Policy. The RACs would contain all relevant stakeholders and be able to offer non-binding advice to the EU and Member States. The implementation of the

MSFD would still take place at the national level but would benefit from best practice knowledge and peer pressure. The latter model has neighbouring Member States working together on an ad hoc basis to coordinate their implementation of the MSFD. This arrangement is largely restricted to information sharing and each country consults with its own stakeholder groups. A more ambitious model the "Regional Sea Convention-Plus" (RSC+) would aim to provide the existing European Regional Sea Commissions with a stronger mandate. Existing national MSFD implementation would be replaced by a regional scale effort coordinated by the RSC+ which holds binding decision making power. Stakeholder consultation would still be at the national level. Finally the fourth model, "Regional Sea Assembly" (RSA) would mark a radical departure from current governance with the Assembly given the power to manage marine regions (Regional Seas) their natural resources, habitats and ecosystem services. The assembly would be composed of elected stakeholder representatives and would implement the MSFD but also regulate other marine based activities.

Van Takenhove et al. (2014) assessed each of the models on the basis of a set of criteria covering cost effectiveness, degree of cooperation, clarity in terms of institutional responsibility and scale mismatch. Their results show not surprisingly that the RSA model scores high on levels of cooperation and on effective institutional arrangements. It is however the most costly model to set up and run. Hybrid approaches may offer the most practicable starting point for governance reform, for example, a combination of AA and RSC+ offers the prospect of effective stakeholder involvement and binding decision making power for a moderate cost burden.

## 1.3 Outline

Following this **Introduction** written from an interdisciplinary perspective, the book continues with a **'Principles' section** with chapters on a conceptual framework for adaptive management; one on a review of existing coastal and marine science models and their strengths and limitations; and one on the methods for valuing ecosystem services. The following section deals with **Practice** and contains chapters on: indicators of ecosystem services change; a review of economic valuation studies; tools for metadata assessment; marine futures scenario analysis. The third section of the book contains **policy relevant case studies**: marine protected areas; managed realignment of coasts; accounting for 'blue carbon'; and blooms of jellyfish.

## References

Barbier, E., Moreno-Mateos, D., Rogers, A., Aronson, J., Pendleton, L., Danovaro, R., Henry, L. A., Morato, T., Ardron, J., & Van Dover, C. (2014). Protect the deep sea. *Nature, 505*(7484), 475–477.

Baveye, P. C., Baveye, J., & Gowdy, J. (2013). Monetary valuation of ecosystem services: It matters to get the timeline right. *Ecological Economics, 95*, 231–235.

Boyes, S. J., & Elliott, M. (2015). The excessive complexity of national marine governance systems – Has this decreased in England since the introduction of the Marine and Coastal Access Act 2009? *Marine Policy, 51*, 57–65.

Crossland, C. J., Kremer, H. H., Lindeboom, H. J., Marshall Crossland, J. I., & Le Tissier, M. D. A. (2005). *Coastal fluxes in the anthropocene*. Berlin: Springer.

Elliott, M., & Whitfield, M. (2011). Chapter 1.07: ecosystem and biotic classifications of estuaries and coasts. In C. Simenstad, & T. Yanagi (Eds.), Volume 1, Classification of Estuarine and Nearshore Coastal Ecosystems (pp. 99–124). In E. Wolanski, & D. S. McLusky (Eds.), *Treatise on Estuarine & Coastal Science*. Amsterdam: Elsevier.

Halpern, B. S., Klein, C. J., Brown, C. J., Beger, M., Grantham, H. S., Mangubhai, S., Ruckelshaus, M., Tulloch, V. J., Watts, M., White, C., & Possingham, H. P. (2013). Achieving the triple bottom line in the face of inherent trade-offs among social equity, economic return, and conservation. *Proceedings of the National Academy of Sciences, 110*(15), 6229–6234.

Mee, L. D. (2005). Assessment and monitoring requirements for the adaptive management of Europe's regional seas. In J. Vermaat, L. Bouwer, R. K. Turner, & W. Salomons (Eds.), *Managing European coasts* (pp. 227–237). Berlin: Springer.

Mee, L. D. (2012). Between the devil and the deep blue sea: The coastal zone in an era of globalisation. *Estuarine, Coastal and Shelf Science, 96*, 1–8.

Norgaard, R. B. (2010). Ecosystem services: From eye-opening metaphor to complexity blinder. *Ecological Economics, 69*(6), 1219–1227.

Raakjaer, J., Leeuwen, J. V., Tatenhove, J. V., & Hadjimichael, M. (2014). Ecosystem-based marine management in European regional seas calls for nested governance structures and coordination – A policy brief. *Marine Policy, 50*, 373–381.

Sagoff, M. (1988). *The economy of the earth*. Cambridge, UK: Cambridge University Press.

Turner, R. K. (1999). The place of economic values in environmental valuation. In I. J. Bateman & K. G. Willis (Eds.), *Valuing environmental preferences* (pp. 2–41). Oxford: Oxford University Press.

Turner, R. K. T., & Welters, R. (2014). *New tools to improve the management of the coastal environment*. Living With Environmental Change, Policy and Practice Notes, Note No. 10, June 2014.

Van Tatenhove, J., Raakjær, J., van Leeuwen, J., & van Hoof, L. (2014). Regional cooperation for European seas: Governance models in support of the implementation of the MSFD. *Marine Policy, 50*, 364–372.

# Part I
# Principles

# Chapter 2
# Conceptual Framework

**R.K. Turner, Marije Schaafsma, Laurence Mee, Michael Elliott, Daryl Burdon, Jonathan P. Atkins, and Tim Jickells**

## 2.1 Conceptual Framework

Coastal zones and their supporting ecosystems present policy makers with a number of challenges including the need to be flexible in the face of dynamic environmental changes, and a high degree of uncertainty about the consequences for ecosystems and socio-economic systems stimulated by pressures and drivers such as, for example, climate change. In this volume the UK NEA ecosystem services framework (ESF) and related decision support tools (see Fig. 2.1) are used as the basis for adaptive coastal management. This strategy is in line with the broadly based ecosystem approach and is now under test or are being implemented across environmental policy circles (e.g., Saunders et al. 2010 for the Crown Estate, Fletcher et al. 2012 for Natural England).

---

R.K. Turner (✉)
School of Environmental Sciences, University of East Anglia, Norwich, UK
e-mail: r.k.turner@uea.ac.uk

M. Schaafsma
Geography and Environment and Centre for Biological Sciences, University of Southampton, University Road, Southampton SO17 1BJ, UK

L. Mee
Scottish Association for Marine Sciences (SAMS), Oban, Argyll PA37 1QA, UK

M. Elliott • D. Burdon
Institute of Estuarine & Coastal Studies, University of Hull, Hull HU6 7RX, UK

J.P. Atkins
Hull University Business School, University of Hull, Hull HU6 7RX, UK

T. Jickells
Centre for Ocean and Atmospheric Sciences (COAS), School of Environmental Sciences, University of East Anglia, Norwich, UK

 
R.K. Turner, M. Schaafsma (eds.), *Coastal Zones Ecosystem Services*, Studies in Ecological Economics 9, DOI 10.1007/978-3-319-17214-9_2

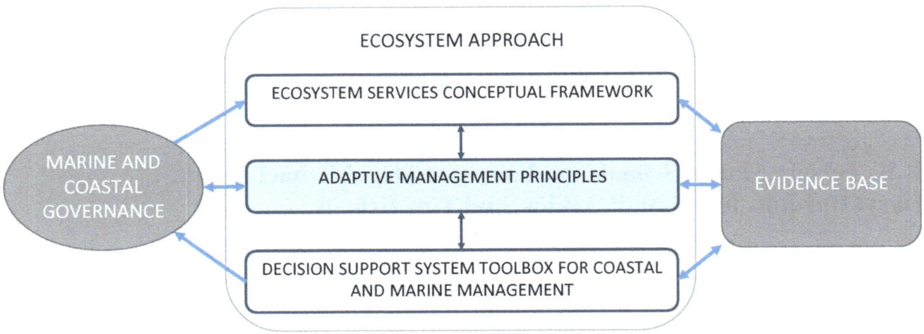


**Fig. 2.1** Implementation of the ecosystem approach (Adapted from the UK NEAFO synthesis report (2014))

A number of flexible 'ground rules' may prove useful in order to guide the application of this ESF and related decision support system (DSS), as well as the interpretation and use of its results by the policy community and society at large. The over-arching adaptive management (AM) approach taken here is built on the foundation principles of pragmatism, decision making anchored to the precautionary principle and pluralism. A pragmatic stance is taken in order to bring the ecosystem services concept more fully into the collective consciousness of government (particularly finance ministries) and business. The methodology therefore deliberately allows for the monetary valuation of the outcomes from 'final' ecosystem services. This stance was pushed further, given the precautionary principle, in the sense that it was judged that sufficient scientific and socio-economic information exists to justify starting to explicitly manage our ecosystems more sustainably and that there is a net benefit from such action. At the same time due recognition needs to be given to the danger of threshold effects because of the scientific uncertainty which shrouds how certain ecosystems may be adversely affected by human development pressures causing them to unexpectedly collapse or lose significant productivity potential.

It will therefore be argued that the ESF also necessitates a plural, interdisciplinary perspective and will require decision makers to operate under conditions of uncertainty, where in some contexts 'full' information will not be available but urgent, or at least short run, precautionary action is necessary. Application of this strategy to dynamic coastal environments and their management will involve just such uncertain and often highly contested ('wicked') policy contexts. Coastal process and ecosystem changes can therefore only be better understood and adaptively managed on the basis of an interdisciplinary 'knowledge' and 'methods and tools' (DSS) capacity.

The coastal management framework set out below is hierarchically arranged. It begins with an explanation of the adaptive management strategy and its high level principles. These were used as guidelines for the deployment of the UK NEA (2011) and UK NEAFO (2014) ESF which in turn provides the focus for a practical DSS, the components of which form the basis for economic and social appraisal and trade-off analysis.

The rest of this chapter is organised into the following sub-sections:

- a characterisation of the strategic-level adaptive management approach encompassing the NEA ESF and the links to relevant decision support tools and methods necessary for more integrated coastal management;
- a classification of coastal and marine ecosystem services, the stock and flow position and the distinction between intermediate and final services;
- the links between processes, ecosystem services and the goods and benefits they provide to human society with wellbeing consequences; and
- an outline of the necessary DSS and its components for practical coastal management.

## 2.2 Policy Context

The interdisciplinary conceptual framework guided by adaptive management (AM) principles and incorporating the ESF and a DSS, seeks to contribute to a more sustainable management of our coastal zones, while *inter alia* at least maintaining the provision of a set of ecosystem services over time. It will also contribute to the UK and other European countries' adoption of the EU Marine Strategy Framework Directive (MSFD) and will draw lessons from the implementation of the EU Water Framework Directive (WFD) and other related Directives and policies, such as the Common Fisheries Policy (CFP). In the UK, for example, the regional marine planning agenda is now the focus of much policy attention driven by legislation such as the UK Marine and Coastal Access Act (2009) and Marine (Scotland) Act 2010, guided by the Marine Policy Statement (MPS) and operationalised by Marine Plans, which set out how the MPS will be implemented in specific areas. The conceptual approach will build on that formulated by the UK National Ecosystem Assessment (UK NEA 2011; Balmford et al. 2011; Bateman et al. 2011) (see Fig. 2.2), and is applied to the coastal zone context. The UK NEA 2011 focused on the processes that link human society and wellbeing to the natural environment and *inter alia* on the key role ecosystems play in delivering a diverse set of services which directly and indirectly underpin economic progress and human wellbeing. The NEAFO (2014) further developed the approach and gave governance and institutions a more central role.

The strategic goal is to build a robust evidence-based case for the embedding of the ESF into the policy process and the workings of the wider contemporary society. However, to foster such a policy switch in practice, new and existing policy tools will need to be combined in a DSS, see Fig. 2.3.

The achievement of the strategic goals of AM will contribute to a better assessment of the value and significance of the flow of ecosystem services over time, as

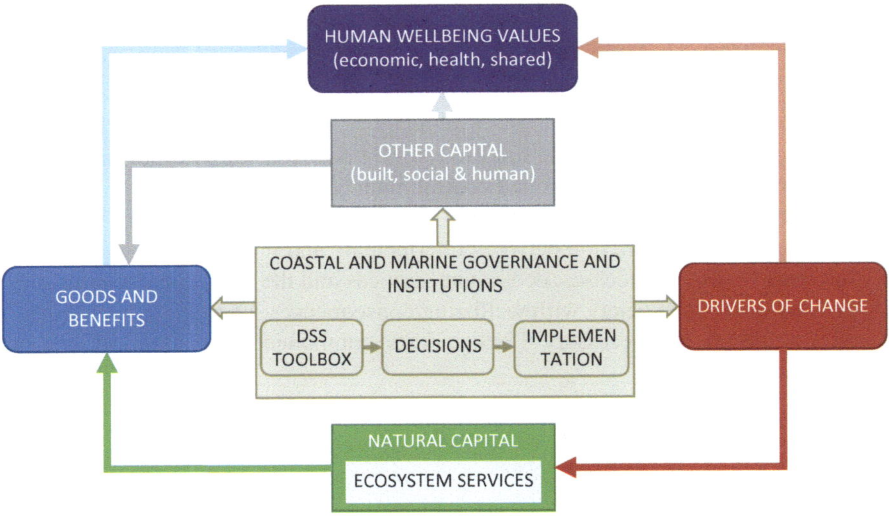


**Fig. 2.2** Ecosystem services conceptual framework (Adapted from UK NEA (2011) and UKNEAFO (2014))

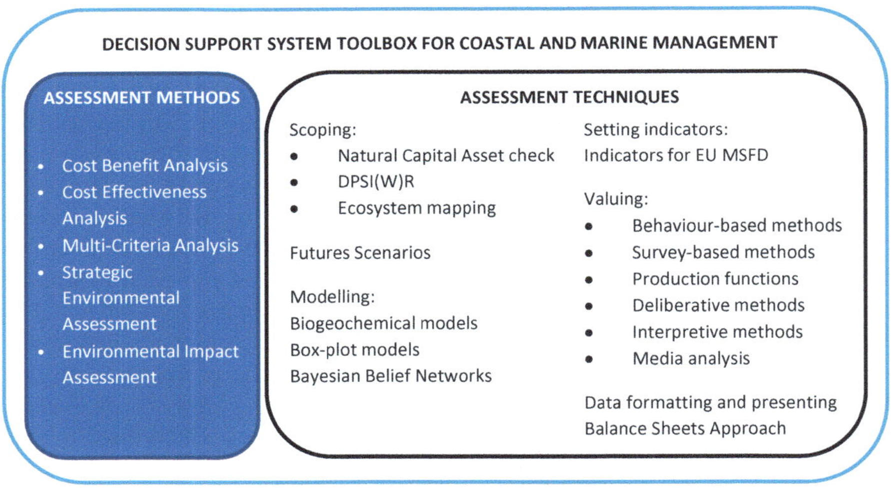


**Fig. 2.3** The DSS toolbox for coastal and marine management with examples of assessment methods and techniques (Adapted from UK NEAFO (2014))

well as an indication of the stock accounting price or value position (natural asset check) at any given point in time. Economic progress cannot be sustainably achieved without good environmental husbandry principles and practice. Sustainability principles can be used to guide the ES framework and approach. This combined approach can then contribute to a fuller quantification and recognition of the true 'comprehensive wealth' of a country (Gross domestic product (GDP) plus) and how it is changing over time (see UNU-IHDP & UNEP 2012). It is also targeted at policy objectives, such as the possible future adoption of a 'strong' sustainable development path (Turner 1993; Turner et al. 2003).

The ESF evolved from an earlier natural science-based analytical approach known as the '**Ecosystem Approach**' as detailed by the 1992 Convention on Biological Diversity (CBD). This advocated a much more comprehensive and integrated approach to environmental management. The next step was to augment the systems-based science by the inclusion of social science and humanities thinking, to link ecosystem functioning and its outcomes to the provision of services (e.g. flood protection, recreation, cultural services and many others) which contribute to human wellbeing. Hence the underlying aim is not so much to solely maximise environmental or biodiversity conservation, but rather to manage the rate of change in ecosystems (structure (including species composition) and functioning (as rate processes)) as socio-economic and ecological systems co-evolve through time.

## 2.3 Adaptive Coastal Management: Principles

Coastal zones are institutional domains with administrative boundaries that can cross regional and national jurisdictions and which are not coincident with the scales and susceptibility of biogeochemical and physical processes (known as the scale mismatch problem). The governance regimes operating across coastal zones therefore face particular challenges. However, political, institutional and coastal management agencies and practices (governance) have so far moved only slowly to encapsulate some core conceptual advances provided by coastal zone 'science' (Mee 2012). These are:

- a recognition that humans are an integral component of the ecology and functioning of ecosystems, and that a process of co-evolution between human society and economy and the environment has now become self-evident due to the scale and intensity of global development and trade;
- environmental management interventions need to be multifunctional rather than focused on single ecosystems or services, with the longer term aim of understanding and managing 'landscape' level ecological processes and relevant socio-economic driving forces and pressures which reduce resilience; therefore the connectivity of a river basin catchment and its receiving coastal waters through to the shelf break is an appropriate functional unit for coastal resource assessment and management;
- to quantify gains and losses from any given policy option choice, it is necessary to assign monetary values to some ecosystem services once translated to societal benefits and to provide non-monetary evaluation of other (particularly cultural) services benefits;

- that new DSSs need to be flexible, allowing refinement and adaptation to changing coastal zone circumstances (such as for example the new focus on marine spatial planning) and governance regimes;
- that some global change impacts (in the absence of radical institutional change at the international governance level) such as temperature change, relative sea-level rise and ocean acidification require a pragmatic adaptive response in advance of long term mitigation and/or compensation;
- because adaptation often results in winners and losers, there is an increasing need for novel forms of compensation in cases where mitigation of adverse effects is insufficient and where the compensation can be for the habitat (e.g. create new habitat), for a resource (such as restocking of affected fish and shellfish stocks) and for users (financial compensation) (Elliott et al. 2007); and
- that the role of the citizen and individual, now often organised via social networks and media, needs to be combined with central decision making in protecting coastal systems quality while at the same time seeking to ameliorate contested values conflict (Potts et al. 2011).

These are all formidable challenges and better DSSs are required if they are to be successfully overcome and progress is made towards more adaptive coastal management. The environmental change forces (often global) that dominate the zone pose risks that are sometimes exacerbated by overly narrow and short term planning and intervention measures, implemented without due regard for ecosystem processes. This temporal mismatch problem is highlighted by situations in which the slow response time of natural systems is challenging for political processes where there are expectations of rapid outcomes from policy interventions. The slow response time also has profound implications for coastal management options and strategies, forcing policymakers to think about taking actions now with consequences that stretch out far into the future. Warming of the deep-ocean and sea-level rise related to increased greenhouse gas (GHG) emissions, for example, are very slow processes taking up to 1,000 years: about a third of the carbon dioxide emitted today will still be in the atmosphere after 1,000 years (Stouffer 2012). We revisit this timescale problem in the context of policy appraisal and the economic discounting procedure in Chap. 4.

In light of the characteristics of coastal zones and policy contexts the adoption of an AM approach at a strategic level is recommended because of, among other things, its emphasis on flexibility and 'learning by doing' practice. Management agencies should therefore be precautionary, giving high priority to coastal functional diversity and related ecosystem services, as well as the maintenance of the system's resistance and resilience (see Box 2.1), i.e. its respective ability to cope with and recover from stress and shock (Turner 2000; Elliott et al. 2007; Elliott 2011). This is a 'stock' quality ('ecosystem health') issue and one that is currently under-researched. We do not know enough about 'minimum' levels of stock structure, processing and functioning and the type and levels of stress that systems can cope with without regime change. This will in turn require the adoption of a relatively broad scale perspective, in order to understand and potentially manage 'landscape' level ecological processes and relevant socio-economic driving forces more cost effectively (de Jonge et al. 2012). A systems-based approach is required to help cope with the inevitable uncertainty that afflicts coastal management and is the basis for AM (Mee 2005).

**Box 2.1. Ecosystem Adaptation**

Ecosystem *adaption* to pressure is a complex process. It can occur at the population and species level as well as within trophic networks. Mechanisms are rarely well known in the case of marine ecosystems, and discussion is often conducted in terms of an emergent property, that of system *resilience*. This refers to the extent that the system maintains its integrity as external pressures increase (*resistance*), or regains that integrity when pressures relax (*recovery*). In Fig. 2.4 the provision of services is shown as a function of ecosystem *state* (indicating integrity or health: see Tett et al. 2013). Recovery, however, may involve change in ecosystem condition (sometimes called regime shift), so that restored services are not identical with those before system collapse.

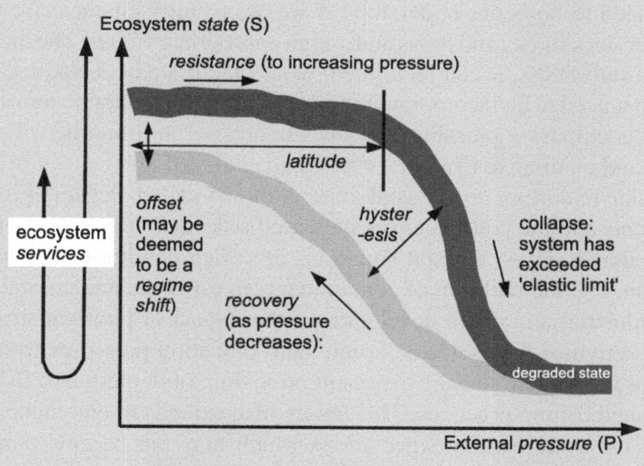


**Fig. 2.4** A conceptual model of changes to the state of a system with increasing pressure (Source: combines ideas in pressure-state diagrams by Tett et al. (2013) and Elliott et al. (2007)

The systems-based approach explicitly recognises that most systems are complex and display inevitable uncertainty in the links between causes and effects. AM is a pragmatic way to achieve national and social-ecological objectives in the face of these high levels of uncertainty. It treats management actions in the coastal and marine system as 'experiments' based on the principle of 'learning by doing'. The MSFD employs this approach through their cycle of target setting, planning, implementation and review of marine strategies (Mee et al. 2008). AM can accommodate 'surprise' events by encouraging approaches that build system resilience to withstand stress and shock and help maintain basic ecosystem functionality (Mee 2005). AM sets both a long term vision (supported by measurable environmental targets, e.g. Good Ecological Status (GEcS) and Good Environmental Status (GEnS) and their indicator sets respectively in the WFD and MSFD), as well as short term goals for ecosystem improvement (see Fig. 1.2). In the case of the MSFD, the long term objectives are supranational (regional sea or EU-wide level), whereas the short-term goals are set through national planning processes and function like 'stepping stones' towards the longer term ones. For 'learning' to occur, it is important that appropriate indicators are formulated (see Chap. 5) and progress towards all targets is monitored carefully and communicated in a transparent manner, allowing objectives and goals to be adjusted from time to time as more information becomes available. The overall vision (GEnS in the case of the MSFD) reflects human values towards the marine environment; the term 'Good' is a human-centric one and the measurement of value is critically important (Mee et al. 2008; Borja et al. 2013).

The linkages between catchment-coastal processes and systems, the influence of climatic change and the impacts on and feedback effects from socio-economic activity all need to be better understood if we are to fully characterise the coastal ecosystem services stocks and flows and assign appropriate values. The incorporation of these data into DSSs, it can be argued, would facilitate better policy outcomes. The values that need to be incorporated are not confined to economic monetary-based values, but encompass a plurality of values expressed in a number of ways, both quantitative and qualitative (Turner 1999; Chan et al. 2012).

A particular feature of the coastal zone is the so-called 'legacy' problem with 'lock-in' effects and the consequential increased risks and vulnerability to flooding and erosion that it poses. Coastal situations are often conditioned by a historical legacy burden, e.g. the build-up of contaminants in estuarine and coastal sediments from past industrial and urban development; the impact of physical structures and reclamation activities themselves; chronic eutrophication pressures from intensive agriculture or inadequate sewage treatment provision; or depletion of fish stocks by long established fishing practices. This legacy also extends to entrenched historical and cultural use patterns and expectations which may not be environmentally or economically sustainable but can be difficult to alter. Thus the impacts on the stock and flow of ecosystem services can be significant, complex and difficult, and costly to ameliorate, often requiring catchment or wider scale action, combined with continual stakeholder engagement.

Social and economic parameters also change as the process of globalisation continues and its pace of change escalates. Driven by the trends in international trade and finance (and fuelled by, among other factors, persuasive advertising industries) coastal zones are at the forefront of a whole suite of continuously evolving impacts with extensive and significant environmental consequences, e.g. from loss of valuable habitats due to port and navigation channel enlargement and energy resource exploitation, to fishing pressures and tourism over-crowding (Mee 2012). Given the plethora of drivers across different spatial and temporal scales, any DSS must be anchored to a systematic scoping process and be tempered by a 'learning by doing' management philosophy that is flexible enough to redo analyses if expectations are not met (Mee 2005). The ultimate goal is to achieve a sustainable and productive utilisation of the available resource system (stock and ecosystem services flow) and the avoidance of irreversible system changes or collapse with consequent high human welfare losses.

## 2.4 Ecosystems Processes and Services: Concepts

Coastal ecosystem natural capital **stocks** (the ecosystem structure and processes and links to the abiotic environment) possess high biological productivity and provide a diverse set of habitats and species, with a consequent **flow** of ecosystem services (the outcomes from the functioning of ecosystems) of significant **value** (benefits) to human society (Barbier et al. 2008). From this valuation perspective, a combination of basic ecosystem structure, processes and 'intermediate' services provide 'final' services of relevance to human welfare ('benefits') as goods that are consumed by humans or essential for human survival (MEA 2005). Ecosystem services benefits are the 'exports' from the ecosystem sector to the human economic sector (Banzhaf and Boyd 2012). Complementary assets (e.g. time, energy, finance or skills) also usually have to be combined with the natural capital to yield benefits. Following the UK NEA (2011) and UK NEAFO (2014) conceptual framework for ecosystem services assessment, the outcomes from the functioning of ecosystems have been generically labelled 'goods' which refer to a range of human welfare benefits derived from the flow of final services provided. But the scope of the delivered final ecosystem services (and therefore the valued goods and benefits) is very wide from food to carbon storage, coastal protection, sea defence, tourism and nature watching (Balmford et al. 2011; Bateman et al. 2011). Figure 2.5 illustrates the conceptual framework begining with boundary conditions. It makes a clear distinction between stocks and flows, and between basic processes, intermediate and final services, and it introduces that values are bounded by beneficiaries groups.

Many definitions and classification schemes for ecosystem services exist (Costanza et al. 1997; Daily 1997; Boyd and Banzhaf 2007). One of the most widely cited is the Millennium Ecosystem Assessment definition (MEA 2005), which describes ecosystem services as 'the benefits that people obtain from ecosystems'. It classifies ecosystem services into: *supporting* services (e.g. nutrient cycling, soil formation, primary production), *regulating* services (e.g. climate regulation, flood regulation, water purification), *provisioning* services (e.g., food, fresh water), and *cultural* services (e.g. aesthetic, spiritual, recreational and other non-material benefits). This framework provides a platform for moving towards a more operational classification system which explicitly links changes in ecosystem services to changes in human welfare. By adapting and re-orienting this definition it can be better suited to the purpose at hand, with little loss of functionality. Wallace (2007), for example, has focused on land management, while Boyd and Banzhaf (2007) and Mäler et al. (2009) take national income accounting as their policy context.

For economic and social valuation purposes the definition proposed by Fisher et al. (2009) clarifies the distinction between ecosystem services and benefits: *ecosystem services are the aspects of ecosystems utilised (actively or passively) to produce human well-being*. Fisher et al. (2009) see ecosystem services as the link between ecosystems and things that humans benefit from, not the benefits themselves. Ecosystem services include ecosystem organisation or structure

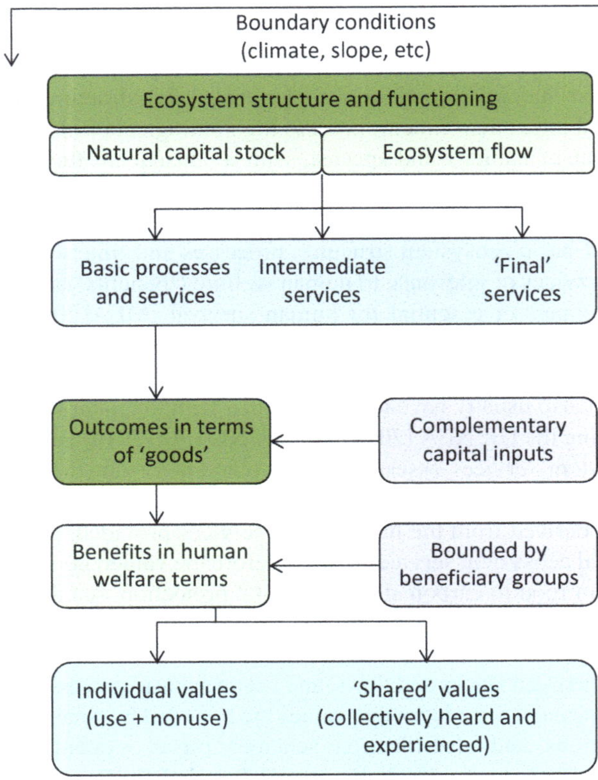


**Fig. 2.5** Ecosystem services conceptual framework

(the ecosystem classes) as well as ecosystem processes and functions (the way in which the ecosystem operates). The processes and functions become services only if there are humans that (directly or indirectly) benefit from them. In other words, ecosystem services are the ecological phenomena, and the good (benefit) is the realisation of the direct impact on human welfare. The key feature of this definition is the separation of ecosystem processes and functions into intermediate and final services, with the latter yielding welfare benefits (see Fig. 2.5).

The term 'intermediate services' should not be interpreted as signifying lesser significance but rather as a necessary signal in order to clearly demarcate (in valuation terms) final services and provide technically-correct guidance to avoid double counting when services are valued in economic or non-monetary terms (Fisher et al. 2009). It is changes in the provision of final ecosystem services that we are interested in measuring and incorporating into economic and social analysis.

The assessment and valuation of ecosystem stock and flow situations is therefore not a straightforward task. The monetary valuation of stocks and flows in particular is complex and has to rely on a range of accounting and socio-economic approaches, together with an underlying natural science understanding (see Chap. 4). Some services will not be amenable to monetary valuation, and the use of coastal resources and their conservation is often highly contested involving different interest groups. Coastal areas are also socio-cultural entities, with specific historical conditions and symbolic significance. The values expressed for such cultural entities may well manifest themselves through collective social networks such as groups, communities and even nations. They may not be best identified through an individual's monetary valuation, but through group deliberation and shared values in quantitative or qualitative terms, or through other evidence sources, e.g. archives (UK NEAFO 2014). We take a closer look at 'shared values' in Chap. 4.

## 2.5 Coastal Ecosystem Processes and Ecosystem Services: Classification

A classification of coastal and marine ecosystem services is provided in Fig. 2.6, whilst Table 2.1 provides a set of definitions supporting this classification, adapted from, inter alia, de Groot et al. (2010), Böhnke-Henrichs et al. (2013), Hattam et al. (2015) and the UK NEA (2011, 2014). The categories and definitions are part of an active research topic and therefore in a process of on-going refinement and improvement. For example, some of the definitions of the cultural ecosystem services in Table 2.1 may overlap, as indicated by footnotes. The overlap should be considered in the assessment of net impacts on well-being.

The set of definitions focuses on those ecosystem services that relate to coastal and marine (C&M) biota, in some cases supported by or dependent on abiotic processes or structures of the ecosystem. Coastal and marine (C&M) biota refers to all living components of the coastal and marine environment including all flora, fauna, algae, bacteria, etc. However, the classification excludes goods and services derived from the abiotic and physic-chemical environment such as the provision of materials for mining, e.g. minerals, oil, marine aggregates, etc.

The definitions support the ecosystem services categories in Fig. 2.6 and were developed for the assessment of (dis-)benefits set in a CBA framework. However, the use of the categories may be adapted to different policy contexts when there is a pre-defined policy objective that society has agreed to. For example, as indicated by a star (*), the policy appraisal concerns a cost-effectiveness analysis and the aim is to manage the coastal ecosystem towards that objective at the lowest possible cost; the emphasis shifts to understanding the ecosystem functioning that supports the policy. If the policy appraisal aims to assess whether costs of measures are (dis-) proportionate, we move back into a CBA framework, and the final ecosystem services related to the standard as well as potential other co-benefits have to be assessed.

Coastal and marine ecosystems are dynamic systems made up of living and non-living components that interact with each other by way of complex exchanges of energy, nutrients and wastes. These exchanges are driven by the physical, chemical

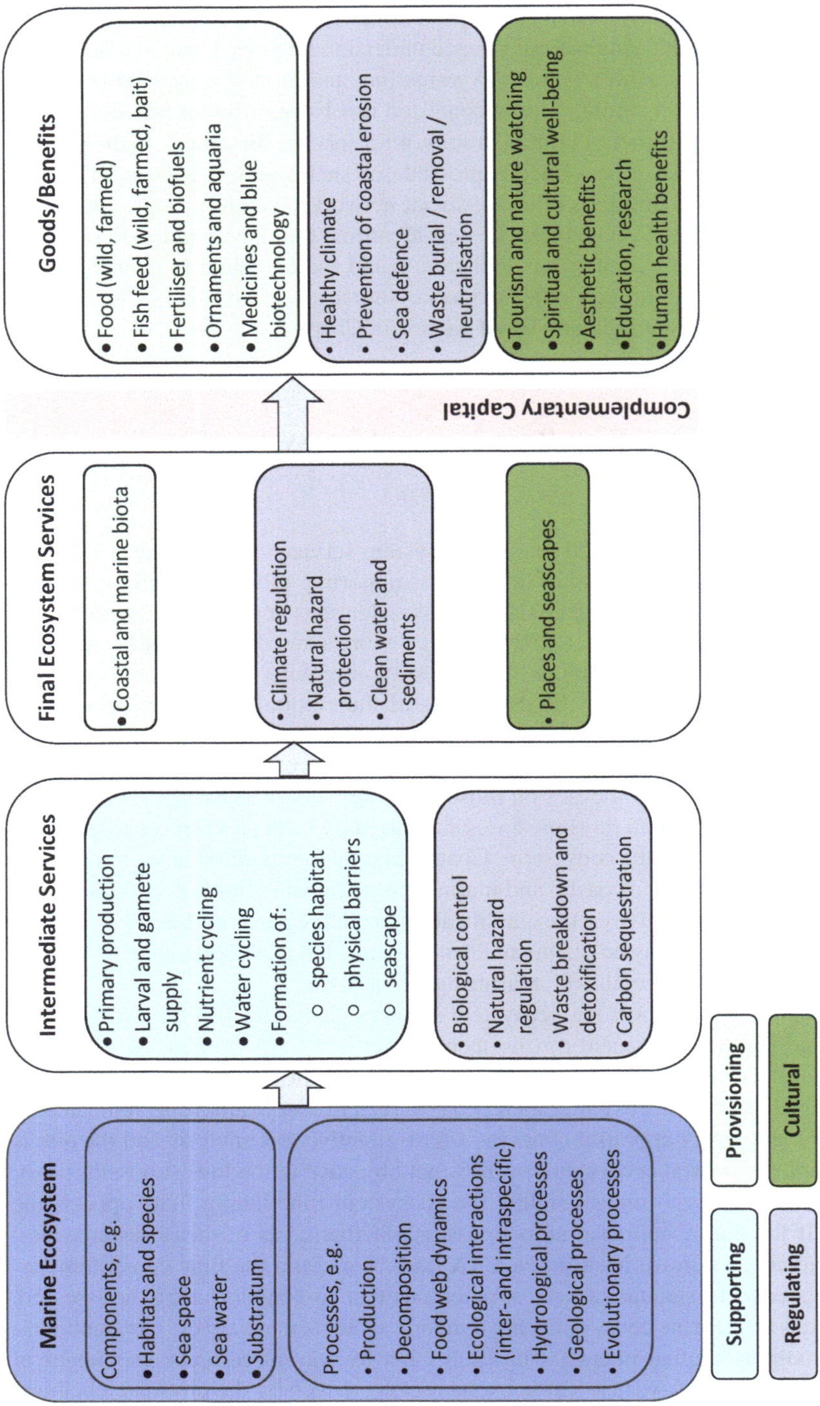


**Fig. 2.6** Ecosystem service classification

**Table 2.1** Definitions of intermediate and final ecosystem services and associated goods and benefits

| | Definition | Example |
|---|---|---|
| **Intermediate services** – *those services whose ecological processes and functions support all life, and, by definition all other services* (UKNEA 2011). *They may include non-fundamental ecosystem processes and functions.* | | |
| Primary production | The synthesis of organic matter by C&M biota from atmospheric or aqueous carbon dioxide | Quantity and/or quality of primary production from a given area of saltmarsh or volume of seawater |
| Larval and gamete supply | The production and supply of larvae and gametes from C&M biota | Quantity and/or quality of larvae or gametes supplied to a given coastal or marine location |
| Nutrient cycling | The influence of C&M biota on the movement or exchange of organic and inorganic matter | Change in the concentration of nitrates/phosphates in coastal or marine waters/sediments |
| Water cycling | The influence of C&M biota on the movement or exchange of water between the C&M environment and adjacent environments (including the atmosphere) | Change in the amount of water retained within a coastal saltmarsh or reedbed |
| Formation of species-habitat | The contribution of C&M biota to habitat formed by one species but providing suitable niches for other species | Change in the formation of mussel beds, kelp forests, cold-water coral reefs |
| Formation of physical barriers | The contribution of C&M biota to the formation of physical barriers | Changes in reef extent by reef-forming organisms (e.g. *Sabellaria spp.*), impacting on the local hydrographic regime |
| Formation of seascape | The contribution of C&M biota to supporting the formation of different coastal and marine views ('seascapes') | Changes in area per type of seascape e.g. algae-covered rocky shore, kelp forest |
| Biological control | The contribution of C&M biota to the maintenance of population dynamics, resilience through food web dynamics, disease and pest control | Oystercatchers controlling intertidal cockle population numbers; cleaner fish (e.g. ballan wrasse) removing sea lice from salmon |
| Natural hazard regulation | The area of suitable C&M habitat which is available to absorb energy | Width or area of saltmarsh/mudflat/reedbed/sea grass |
| Waste breakdown and detoxification | The presence of C&M biota which have the potential to remove anthropogenic contaminants and organic inputs | The presence of reedbeds, mussels beds, etc. |
| Carbon sequestration | The net capture of carbon dioxide by C&M biota | Change in the net amount of carbon stored within an area of coastal saltmarsh within a certain period |

(continued)

**Table 2.1** (continued)

| | Definition | Example |
|---|---|---|
| **Final services** – *the outcomes from ecosystems that directly lead to good(s) that are valued by people* (UKNEA 2011) | | |
| C&M biota | The flow of C&M biota | Change in the quantity/quality of North Sea cod population, seaweed stock, genetic material, ornamental materials, etc. over time |
| Climate regulation | The contribution of C&M biota to the maintenance of a favourable climate through the regulation of greenhouse gases | Healthy climate |
| Natural hazard protection | The contribution of C&M biota to the dampening of the intensity of environmental disturbances such as storms, flooding and erosion | The reduction in the intensity of environmental disturbances resulting directly from C&M ecosystem structures such as saltmarsh and sea grass beds |
| Clean water and sediments | The contribution of C&M biota to the provision of clean water and sediments | Quantity of waste (tonnes) that is recycled or immobilised by C&M biota over a period of time |
| Places and seascapes | The contribution of C&M biota to places and seascapes | Number of coastal sites designated for internationally important seabird colonies |
| **Goods/benefits** – *all use and non-use, material and non-material outputs from ecosystems that have value for people* (UKNEAFO 2014) | | |
| Food (wild, farmed) | Extraction of C&M biota for human consumption | Tonnes of cod landed for human consumption |
| Fish feed (wild, farmed, bait) | Extraction of C&M biota for non-human consumption | Tonnes of sandeel harvested to be processed into fishmeal; volume of mackerel caught for use as bait in crab/lobster pots |
| Fertiliser and biofuels | Fertiliser (biocides) or energy sourced from C&M biota | Biomass of algae harvested to be processed into fertiliser |
| Ornaments and aquaria | Extraction of C&M biota for decoration, fashion, handicraft, souvenirs etc. or for display in aquaria | Number of European lobster extracted for display in aquarium exhibits; amount of skins, shells, corals, plants, extracted from the C&M environment for decoration, fashion etc. |
| Medicines and blue biotechnology | Extraction of C&M biota in order to produce medicines, pharmaceuticals, animal and plant breeding and biotechnology | Marine-derived pharmaceuticals such as the use of sea lettuce (*Ulva lactuca*) in cosmetic and personal care items including make-up remover, shampoo and shaving lotion |
| Healthy climate | Improvements to human well-being as a result of a healthy climate | Bodily harm avoided as a result of natural carbon sequestration by C&M biota |

| | | |
|---|---|---|
| Prevention of coastal erosion | Reduction in hazards resulting from the natural prevention of coastal erosion by C&M biota | Prevention of gradual damage to property and land by dunes |
| Sea defence | Reduction in flooding related hazards as a result of the natural protection provided by C&M biota | Saltmarsh providing a natural form of sea defence in the coastal region |
| Waste burial/removal/ neutralisation (*) | Contribution of C&M biota to achieving pre-defined policy standard related to waste levels in water by natural waste burial, removal and neutralisation | Natural waste breakdown by C&M biota such as reedbeds – in contexts in which pre-defined regulations/ standards apply |
| Tourism and nature watching | Benefits from recreation, leisure driven by coastal seascapes and their associated C&M biota | Benefits associated with watching seabirds, marine mammals |
| Spiritual and cultural wellbeing[a] | Ability to enjoy preferred lifestyle, culture, heritage, folklore, religion, creative inspiration, and spirituality; sense of place (use-driven) based on ecosystem aspects | The importance of C&M environments in cultural traditions (e.g. traditional cobble fisheries on east coast) or folklore (e.g. sea shanties) |
| Aesthetic benefits[b] | Enjoyment of the beauty of C&M seascapes | Higher house prices in coastal locations |
| Education, research[c] | Enjoyment of formal and informal education, research and science, knowledge systems, etc. in which C&M biota play a role and are a source of information | Amount of funding secured for research on C&M biota; number of scientific research papers published which focus on C&M biota |
| Health benefits[d] | Human physical and psychological health benefits associated with the direct and indirect use of the coastal and marine environment | Increased psychological well-being from direct or indirect experience of the C&M environment; increased physical well-being resulting from engagement with C&M environment, e.g. exercise. |

[a]Overlap with recreation and human health related goods and benefits categories should be checked and where possible avoided in the valuation assessment. Similarly, there is some ambiguity in the distinction between art and design and aesthetic benefits

[b]Aesthetic benefits may also be reflected in tourism and nature watching and ornamental values

[c]Overlap with the category medicines and blue biotechnology should be avoided in the valuation assessment

[d]Overlap with tourism and nature watching, spiritual and cultural wellbeing, medicines and food

and biological processes or attributes that are characteristics of a particular ecosystem and its functioning. The functioning of coastal and related marine areas is maintained through a diversity of ecosystems, e.g. salt marshes and other wetlands, sea grasses and sea weed beds, beaches and sand dunes, and estuaries and lagoons. This natural capital stock provides a range of processes such as nutrient and sediment storage, water flow regulation and quality control and storm and erosion buffering (see Fig. 2.6.) (Crossland et al. 2005). Coastal and marine ecosystem processes and functions can, for example, be grouped into four broad categories, which broadly map on to the processes, 'intermediate services' and final services concepts in the classification system adopted here to facilitate monetary valuation:

- **Purification and Detoxification:** filtration, purification and detoxification of air, water and soils;
- **Cycling Processes**: nutrient cycling, nitrogen fixation, carbon sequestration and soil formation;
- **Regulation and Stabilisation**: pest and disease control, climate regulation, mitigation of storms and floods, erosion, regulation of rainfall and water supply; and
- **Habitat Provision**: refuge for animals and plants, storehouse for genetic material.

An intermediate service is one which influences human well-being indirectly, whereas a final service contributes directly. Classification is context dependent, for example, clean water supply is a final service to a person requiring drinking water, but it is an intermediate service to a recreational angler. Importantly, a final service is often but not always the same as a benefit. For example, recreation is a benefit to the recreational angler, but the final ecosystem service is the provision of the fish population. This examples shows how the classification approach used here seeks to provide a transparent method for identifying the aspects of ecosystem services which are of direct relevance to economic valuation, and critically, to avoid the problem of double-counting.

The policy context to which the analysis relates is also very important and influences the way in which the ecosystem classification can be utilised. To take an example, an estuary and coupled catchment characterised by, among other economic activities, intensive agricultural regimes. The estuary has extensive wetlands, salt marsh and mudflat areas which can provide a set of ecosystem services. Given the impacts of intensive agriculture, for example, heavy nutrient N and P runoff, the wetlands can provide valuable services such as nutrient cycling. If for example, national policy includes a provision to increase wetland habitat and the services it provides, in a CBA of this policy option the nutrient cycling service provided by the wetlands would be treated as an intermediate service contributing to the provision and value of final services, e.g. better water quality. This cleaner water may then lead to enhanced recreation and amenity benefits, or improved fisheries productivity, which can be assigned a monetary value.

A change in the policy context, however, can change the way in which the ecosystem service classification is used. Assume the estuary is already subject to an official (national or international) water quality standard provision, which it is failing and the policy option under consideration is how best to meet the standard. Now cost effectiveness analysis (CEA) would be deployed to determine the least cost way of achieving the pre-existing water quality standard. In this context the nutrient cycling service provided by an increase in the wetlands via re-creation, would be focused on and the costs of wetland re-creation or establishment would be compared with, for example, the cost of enhanced sewage treatment processes and facilities, or changes in agricultural regimes imposed on farmers (e.g. nitrogen zoning).

## 2.6 Decision Support System (DSS): Practice

The DSS needs to be composed of a number of sequential (depending on the exact policy issues and context) but overlapping components:

- An interdisciplinary scoping exercise to establish or model baseline ecosystem and co-evolving socio-economic systems conditions and trends, together with a focused attempt to identify 'key' policy contexts and issues;
- The selection and development of appropriate functionally related indicators of ecosystem state (the stock position) and changes in services (the flow position) supply over time;
- A futures assessment through the use of scenarios covering prevailing conditions and alternative future states;
- The deployment of 'tools' (including models) to enable a scientific, economic and social appraisal of policy options, including distributional concerns and the use of deliberative methods and techniques to foster social dialogue across interest groups;
- Appropriate formatting and presentation of appraisal data, assumptions and findings into an evidence base; and
- Setting up adequate monitoring and review procedures.

We look at the main components of the DSS below.

## 2.7 Scoping Environmental Change in Coastal Zones

The underlying activity-pressure-impact chain characteristic of coastal zones (Crossland et al. 2005) can be been expanded to form the Drivers, Pressures, State changes, Impacts and Policy response (DPSIR) framework. Further, because of the continuing confusion between the S being State and State Change and the I being Impact (on the natural system) and Impact (on the human system) (Atkins et al. 2011), the original formulation has been further modified to the DPSWR approach where W replaces I as impact on human welfare (Turner et al. 1998; Cooper 2013).

The DPSI(W)R framework can help to scope in a standardised fashion policy and management contexts in order to get a better understanding of this environmental change process and what it means in ecosystem service terms. This established scoping methodology can combine data about environmental change drivers and pressures with causal mechanisms which result in environmental state changes, and impacts associated with human welfare gains and losses. Feedback loops between policy responses and other components of the change process are also encompassed within the approach to avoid overly linear thinking as individual and societal innovation often occurs in a non-linear and in sometimes surprising ways. The approach first developed to classify and organise environmental indicators has proved to be a useful heuristic in wider environmental management contexts (Turner et al. 1998). The scoping exercise has to be sufficiently robust to capture all the main drivers of change and behaviour incentives across multiple actors, jurisdictions and agencies. While it is the case that coastal and marine system issues can be complex and that a range or combination of variables influence human interest individuals and groups, under any given governance system, partial decomposition of problems is possible (Ostrom 2007). However, the information provided by the DPSWR process will require further refinement to include a specific focus on ecosystem services and in order to highlight 'key' contexts and issues. The Impacts or Welfare stage needs to be specifically calibrated in terms of ecosystem services and interactions and feedbacks (Kelble et al. 2013).

The *framing* of a policy issue is necessary in order to enable identification of appropriate decision support processes and suitable policy instruments. Typical contemporary policy issues within the regional seas and coastal zones, and which are at the core of the need for better policy tools and governance regimes are diverse. For example, increase in human population size may lead to increase building activities in risk prone zones, including more artificial defence structures, which in turn can lead to the destruction of natural habitat such as saltmarsh, or arable land. Aquaculture and wind farm development may lead to pollution and loss of habitat and biodiversity, which consequently affects goods and benefits such as fisheries and recreation, either directly or by providing a stepping stone for invasive species.

Figure 2.7 illustrates the DPSWR framework in standard form, including feedback loops between Responses, and Drivers and Pressures, and recognition that there are natural pressures on ecosystems, which can lead to State Changes. Defining boundaries requires due care and attention, because pressures on the system can be locally, regionally or internationally managed pressures (power generation, fisheries, etc.), or exogenic unmanaged pressures (climate change, volcanic eruptions, geomorphic isostatic readjustment, etc.). The latter case, in contrast to the former, is one of bounded rationality (i.e. taking action with limited information on a 'learning-by-doing' basis) since their complexity is such that we do not yet have sufficient knowledge of how and why change occurs in such systems, and so our response is not of the management of the pressure but of the consequences of that pressure; in the case of endogenic managed pressures, we may be able to manage both the causes and the consequences (Atkins et al. 2011).

The DPSWR framework has been widely used to assess and manage the impact of policy changes and associated problems; however, a change is evident in recent

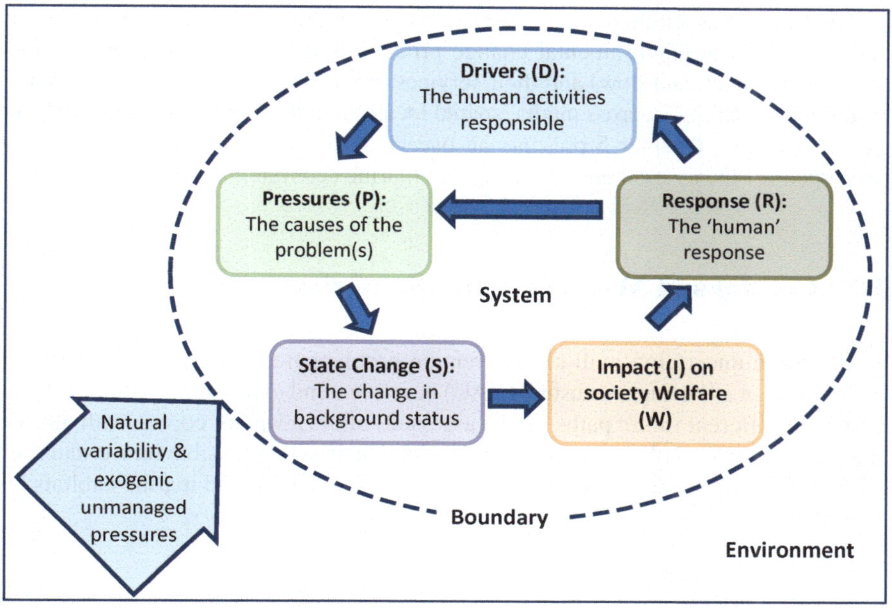


**Fig. 2.7** The DPSI(W)R framework. DPSI(W)R can be explicitly focused on ecosystem services through the S and I(W) stages

applications of the approach: an expert-driven, evidence focused mode of use is giving way to the use of the framework as a heuristic device to facilitate engagement, communication and understanding between different stakeholders (Cooper 2013; Kelble et al. 2013). The DPSWR has been used to categorise indicators of environmental change and the application of scenario analysis to the framework can also be a useful way to further embed the DPSWR into the DSS for management.

## 2.8 Indicators

The future challenge in the EU is the joint implementation of the WFD and MSFD with the former focusing on the protection of the system according to chemical status and five biological quality elements (four in the coastal zone), whereas the MSFD focuses on 11 descriptors, each of which can be linked to show a hierarchy (see Borja et al. 2013). The WFD is regarded as a 'deconstructing structural' approach, whereby the indicators are more easily related to the structural ecosystem components, whereas the MSFD apparently will relate to functioning of the system and a more well-defined set of pressures along the activity-pressure-impact chain (Borja et al. 2013).

The MSFD has stimulated new work into appropriate indicators linked to the 11 descriptors of the environmental change process as it affects coastal and marine ecosystems (stock and flow) and their services provision. Functional indicators are required, for example, across media, spatial location, hydrological function and biological function. Chapter 5 presents an overview of the indicators that are being developed for the assessment of coastal and marine ecosystems.

## 2.9 Coastal and Marine Futures Scenarios

While future uncertainty will always remain problematic, scenario analysis (typically based on a ‘business as usual’ (BAU) baseline trend assessment, against which a range of different future paths can be assessed) offers a way of coping with uncertainty and provides policy relevant decision information on plausible future states of the world. Chapter 8 discusses possible scenarios for coastal and marine habitats in more detail.

## 2.10 Models

An important component of the AM approach and DSS is the development of models. A number of different types of models can be deployed, ranging from formal scientific models of land use change in catchments with links via nutrients and other factors into models for estuaries and coastal waters, to conceptual models which are simple ways of highlighting and eliciting human perceptions about how a system functions. The latter allow a dialogue between experts, stakeholders and the public which conveys information, identifies ‘contested’ issues and provides the opportunity to reinforce or modify perceptions and expressed values (Turner 1999). Underpinning the approach is a requirement to collect empirical data and metadata on ecosystem functioning and service provision, together with an understanding of the distribution of ecosystem benefits (who gains or losses in any environmental change situation) and governance contexts. We review the available models for coastal and marine systems in Chap. 3.

## 2.11 Economic and Social Appraisal

The application of economic and social appraisal of projects, policies, programmes or courses of action in the coastal context can only take place after policy issues have been identified and highlighted within given spatial and temporal scales, and scenarios and evaluative criteria have been established and legitimised within the dialogue process. Once agreed, the policy issues and scenarios chosen then provide the backdrop and framework within economic and social appraisal can take place. However, this is not a one-way process. Ideally, feedback should occur between all stages of the assessment process and the deliberative procedures set up with stakeholders, since concerns that are thrown up by the dialogue can help to refine the policy issues, leading to acceptable interventions and scenarios that resonate with most stakeholders and interest groups.

### *2.11.1 Environmental Impacts, Welfare and Economic Values*

Once policy issues and scenarios are established, the next stage of the process is to determine all the relevant impacts that will take place under the scenarios considered. These impacts relate to changes in the provision of final ecosystem services and goods (which could include, for example, the carbon storage functions of coastal mudflats) and other, more conventional goods (such as commercial fish catch or shellfish harvested from coastal mudflats). Primarily, economic assessments are concerned with those impacts on goods that can be valued in monetary terms. However, this does not mean that all impacts can be incorporated into such an analysis – it may not be possible to value all impacts in this way, because of practical or ethical considerations. Hence we consider that economic assessment provides just one strand of an overall integrated (sustainability) analysis, with other strands being supplied by assessments and techniques from social, deliberative and ecological perspectives (such as multi-criteria analysis (MCA), participatory GIS, deliberative fora, deliberative monetary valuation). It is also the case that the sustainable provision of the flow of final services and related goods and benefits depends on the maintenance of system-wide ecosystem processes with adequate carrying capacity and resistance and resilience characteristics. Conventional economic analysis based on marginal changes is not well suited to identifying and encompassing system unsustainability.

The core of the economic assessment process is to determine how changes in ecosystem services provision are translated into changes in welfare (which can be positive or negative, i.e. benefits or costs). This is achieved by placing a monetary value on each of those changes and aggregating these values together to arrive at an overall change in value for the environmental and policy scenarios considered. Chapter 4 discusses the non-market valuation theory.

### *2.11.2 Policy Response Interventions*

Policy response interventions usually fall into a number of categories:

- **Mitigation of pollution and resource overexploitation problems** – the ecosystem service benefits that need to be valued are related to damage reduction and/

or restoration measures, e.g. reduced flooding damage or sedimentation in navigation channels or restoration of wetlands, water treatment investment, changing farming practices in the catchments, etc.;
- **Compensation for losers measures** – these may be financial as in the case of coastal erosion problems in England and Wales with, for example, the Pathfinder experimental scheme in which local authorities offered to pay 40–50 % of the theoretical value of properties threatened by coastal erosion, based on the value of similar properties inland; or environmental compensation under a precautionary principle, safe minimum standards approach, which can include project management on a portfolio basis (Barbier et al. 1990) with so-called 'shadow' or 'compensating' projects; or habitat equivalency compensation measures (Roach and Wade 2006);
- **Enhancement of marine and coastal zone ecosystem services** – actions which provide an increased provision of benefits, e.g. adaptation to change (see Box 2.1), which increases the output of some good such as creation of artificial reefs to provide erosion protection, or fisheries habitat and nursery which enhance productivity of the stock, or the reduction of conflicts among or between various users of coastal ecosystems via pricing schemes or zoning;
- **Preservation of unique marine and coastal ecosystems** – the benefits stem from setting aside and managing particular areas via Marine Protected Areas (MPAs) in order to preserve the natural ecosystem can be twofold. Use benefits e.g. visits to a nature reserve to observe nature or take photographs, etc.; and non-use benefits which are not related to visits but encompass option or existence values. The non-use values here relate to motivations which seek to conserve ecosystems for future use (insurance value) and the continued presence of species and habitats from which people derive passive welfare. Shared values will also be important in this category; and
- **Joint usage benefits –** within this last category of interventions, marine spatial planning and zoning have recently come to the fore, including the search for joint usage benefits. The UK Marine Policy Statement, for example, contains the following statement: *"The Marine Plan should identify areas of constraint and locations where a range of activities may be accommodated. This will reduce real and potential conflict, maximise compatibility between marine activities and encourage co-existence of multiple users"*.

There is a need to better understand the barriers to the achievement of joint net benefits, i.e. co-location situations in which multiple users or activities share the same impacts footprint (MMO 2013). The decision to locate any given economic activity in a particular marine space will be conditioned by a range of factors. At the core of this process will be an assessment of financial profit or loss potentially available to the economic agent (individual or firm) involved. However, the decision will be further constrained by existing and possible future legislation and regulation and wider social and environmental issues, such as, for example, loss of local employment or cultural identity when fishing activities are curtailed or lost; and environmental impacts including use and non-use loss if biodiversity is reduced. So the impacts (footprint) of co-location can be multidimensional and any assessment method must be able to accommodate this diversity. The Balance Sheets Approach framework set out in Sect. 2.13 seeks to meet this need.

Two economic concepts, externalities and joint production, can be used in order to formally distinguish between the different possible categories of co-location. The 'technological externalities' concept refers to the indirect effect of an economic agent's consumption or production activity on the products, consumption or welfare of a different economic agent, and where the effect does not work through the price system. Externality effects can be positive or negative and quite diverse, including forms of pollution or contamination and interaction between different production activities. In the latter context, so-called 'joint production' cases can be identified. So multiple products may be produced under separate production processes, or several outputs may be produced from a single production process.

Three distinct categories of co-location for a given marine space can be identified using the economic concepts of externalities and joint production (see also Lester et al. 2013):

- No co-location – situations in which there are no feasible joint production possibilities and candidate activities generate negative externality effects; e.g., offshore wind farms and demersal fishing with beam trawls cannot take place at the same location;
- Horizontal co-location – joint production possibilities exist and the candidate activities do not generate significant negative externality effects; e.g., offshore wind farms and open water aquaculture can go together; and
- Vertical co-location – no joint production possibilities and no negative externality effects; e.g., recreational fishing or boating in a MPA but limited to certain times of the year to protect fish spawning or biodiversity.

## 2.12 Balance Sheets Approach Format

Finally, we turn to the question of how appraisal might be sequenced and how information can best be collated, interrogated and presented to policymakers. Building on the work of the UKNEA (2011), the UK NEAFO (2014) has developed the Balance Sheets Approach as a means constructing as robust an evidence base as is feasible to underpin the policy process. It is therefore both a process and a tool and forms one component of an overall DSS.

If CBA or related methods are to continue to play a role in the policy process, then a more explicit focus on distributional issues (i.e. who gains and who loses from environmental change and consequent policy responses) is required. A two stage approach needs to be adopted in which the spread of costs and benefits across different affected individuals and groups in society needs to be accounted for, and a weighting procedure applied. Project appraisals funded by economic development agencies have routinely included distributional weights but this practice has not been common place in other public sector applications. As a minimum, the way in which the CBA 'accounts' are set out and formatted needs to be changed in order to incorporate and highlight financial transfers and the distributional impact of costs and benefits across stakeholders. Krutilla (2005) has set out a tableau format which disaggregates the benefits and costs of a project or policy among stakeholders and records all inter-stakeholder financial transfers. It also serves to illuminate key issues such as the level of aggregation adopted and the project or policy accounting boundary.

Changing the accounts format is a necessary first step, but Kristrom (2005) has gone further and put forward a 'hierarchy of options approach' in which explicit distributional weighting is applied, based on a rule that requires higher weights on all costs and benefits accruing to socially disadvantaged or below average income groups. Alternatively, explicit distributional weights can be introduced to reflect the degree of inequality aversion present in society, by examining past public policy decisions, or the prevailing marginal rates of income tax (Atkinson et al. 2000).

Any DSS that is put in place to assist in evaluating the gains and losses involved in marine planning and management will need to encompass a wide diversity of impacts and different stakeholder perspectives. The Balance Sheets Approach (see Fig. 2.8) is a pragmatic attempt to provide a framework within which the complexity of real world decision making and trade-offs can be examined. It sets out three complementary components (balance sheets) which can be seen as 'roughly comparable' sets of findings with overlaps and linkages. The aim would therefore be to determine the 'best' combination of data, methods and analysis, depending on the actual activity and context under appraisal (Turner 2011). The different policy contexts are illustrated in the figure along the horizontal axis in terms of a spectrum between two polar opposites: slow and simple versus complex and dynamic change processes.

The complexities and the non-commensurate values that characterise the real world political economy of 'contested' natural resource allocation and trade-offs are clearly illustrated in European fisheries policy. The annual fisheries negotiations in Brussels try to set rules for fair access to fish stocks. Scientists have recommended total limits to catch to avoid fishing beyond levels that the stock will support. Ministers then meet together at the annual Fisheries Council to set pragmatic rules of access based on instruments such as gear type, number of vessels, days at sea and total allowable catch. In the past, ministers have often negotiated catch allocations that exceed the advice of their own scientists. One reason for this is the non-commensurability of the currencies used by different sectors engaged in the process, each of which seeks to archive 'sustainability'. The fleet owners seek to sustain profits (market values, that can be subjected to a CBA); local political representatives seek to optimise or conserve employment and multiplier effects at the community level (measured as jobs and susceptible to financial impact analysis at a local scale); and conservationists emphasise non-use values and ethical considerations (more amenable to deliberative methods including MCA). The Minister at the Council tries to balance these interests but, without an effective analytical framework, and with competing claims from other ministers, the likelihood of

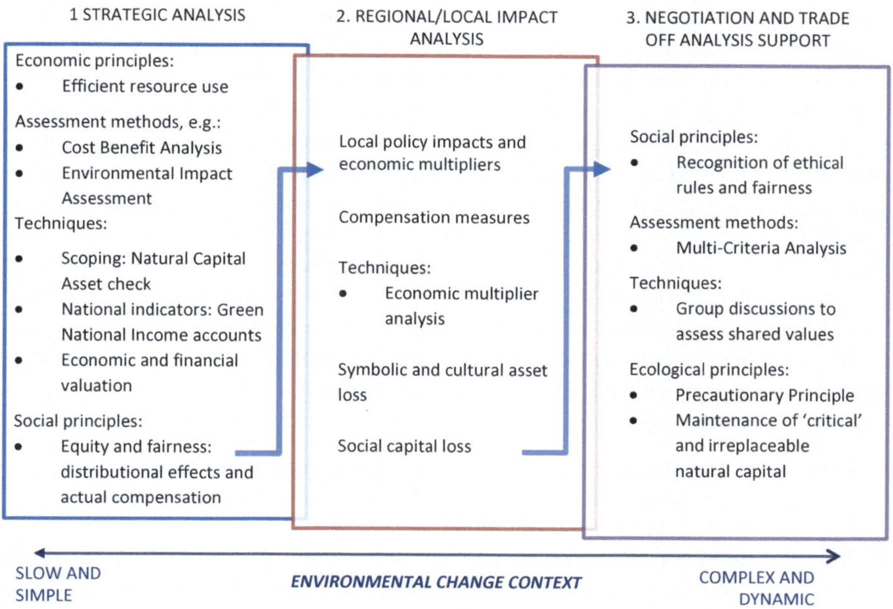


**Fig. 2.8** Balance sheets approach. Assessments progress through the sheets as the social and environmental context become more complex and dynamic

success is quite low. The next reform of the EU Common Fisheries Policy will try to improve this situation by following the 'Ecosystem Approach' that recognises humans as an intrinsic part of the system and that total allowable catch or damage to habitats and non-target species cannot be permitted to exceed ecosystem limits.

Another policy context concerned with coastal protection and sea defence also highlights the 'wicked' characteristics common in many environmental management situations. Over the past decade or so UK government policy in terms of future investments in coastal management has been re-orientated away from a 'hold the line' philosophy and towards a more flexible approach. The new approach has included coastal realignment schemes in selected locations and also a greater recognition of coastal processes such as erosion and subsequent beach replenishment. But the DSSs and policy planning had not been sufficiently adjusted before the headline strategic policy shift became widely publicised and stakeholder concerns were raised. Poor policy support sequencing has meant that difficult 'local' policy impacts and controversies have been raised and policymakers have been slow to respond. Thus the switch towards a more flexible coastal management regime can be justified on overall cost grounds and national strategic requirements, together with a precautionary approach to possible climate related sea level rise and storm intensity and frequency predictions. But the distributional consequences should have been recognised in advance of the policy switch, and mitigation measures should have been in place, as well as a more targeted information and awareness campaign. Instead the agencies involved have had to play catch up, following numerous stakeholder protests and campaigning and wide press coverage. So acceptable 'compensation' measures for the 'losers' in any given coastal scheme (and for that matter flooding risk situations more generally across catchments) have only slowly emerged as controversy has escalated. The pathfinder scheme trialled in East Anglia, England, for example, has examined a number of compensation measures for householders affected by coastal erosion. Under a Balance Sheets Approach the distributional impacts and 'local' impacts would have been diagnosed prior to the strategic policy switch, policy options would have been assessed and arguably more effective ameliorative measures would have been in place.

In the Balance Sheets Approach, three types of complementary assessments (balance sheets) are envisaged to try to give some guidelines for steering a reasonably objective course through these 'contested' policy contexts (see Fig. 2.8):

- Economic (monetary) CBA using a conventional economic efficiency criterion (macro UK economy efficiency), but augmented with a distributional analysis of impacts and possible equity weighting;
- Regional and local financial impacts and policy analysis, covering impacts like local unemployment, loss of community identity and related financial multiplier effects which often raise issues of compensation; and
- Trade-off analysis (non-monetary) better suited to dealing with collective or shared values across wider society such as, for example, intrinsic value in biodiversity, cultural services value etc.

The analytical sequence of the Balance Sheets Approach would typically begin with an economic cost-benefit scoping analysis and then proceed to include the other balance sheets depending on the issue and context under scrutiny. The aim would not be to aggregate the results of each balance sheet, but to present the policy process with the set of findings in as transparent a way as possible.

Given the range of data that relates to the marine environment and related socio-economic activities, there is a pressing need to agree broad categories of data which can illuminate the economic, social and environmental dimensions of environmental change in the marine context. The Balance Sheets Approach aims to achieve this by separating out, in the first instance, economic data and analysis. So in the first column of Fig. 2.8 economic data is covered and is guided by the criterion of macro-economic efficiency and informed by market-based data, willingness to pay (WTP) data and cost data (including second best data such as GVA, etc. – see Chap. 6). A key link to the second column in Fig. 2.8 is provided at the bottom of the first column when the issue of the distribution of costs and benefits is raised, i.e. who gains and who losses from any change. The second column of Fig. 2.8 now expands on the sort of data and issues that are best classified as social effects with a spatial boundary (local to regional) condition imposed on the analysis. The final column continues the social analysis but now encompasses values and impacts that are often expressed at the national scale with a variety of underlying ethical criteria. Clearly the columns overlap, but the aim is to give some logical sequence to a decision support method(s) and processes which are trying to scope and analyse real world (often 'wicked') economic and socio-political issues.

In ideal circumstances, the framework of action to deliver sustainable management needs to fulfil a set of tenets covering all facets of decision making and the identification of defendable sustainable development measures, especially in 'wicked' policy contexts (Elliott 2011, and references therein). These indicate that our actions are required to be environmentally or ecologically sustainable, economically viable, technologically feasible, socially desirable or tolerable, administratively achievable, legally permissible and politically expedient. These seven tenets (Elliott 2011) have been augmented by a further three tenets: ethically defensible (morally desirable), culturally inclusive and effectively communicable (Elliott 2013). This is a formidable list of requirements and pragmatism rather than a futile search for meta-ethical perfection is the recommended course of action under AM.

The final column's information also contains a reminder to set the proceeding analysis in an overall systems context, with due regard for threshold effects and the overarching goal of sustainable development. But following this guidance almost inevitable means trade-off choices and therefore winners and losers. The exact combination of decision criteria and support tools that are relevant will depend on the prevailing and expected policy context and the type of trade-off. The heavy, extensive and on-going utilisation of coastal and marine resources ensures that management decisions will be contested by competing interests. The goal of a return to good (pristine) conditions (Hering et al. 2010) is also unlikely to prove practicable, and so the DSS and social dialogue has to focus on the future and feasible future environmental system states.

## References

Atkins, J. P., Burdon, D., Elliott, M., & Gregory, A. J. (2011). Management of the marine environment: Integrating ecosystem services and societal benefits with the DPSIR framework in a systems approach. *Marine Pollution Bulletin, 62*, 215–226.

Atkinson, G., Machado, F., & Mourato, S. (2000). Balancing competing principles of environmental equity. *Environment and Planning A, 32*, 1791–1806.

Balmford, A., Fisher, B., Green, R. E., Naidoo, R., Strassburg, B., Turner, R. K., & Rodrigues, A. S. L. (2011). Bringing ecosystem services into the real world: An operational framework for assessing the economic consequences of losing wild nature. *Environmental and Resource Economics, 48*, 161–175.

Banzhaf, H. S., & Boyd, J. (2012). The architecture and measurement of an ecosystem services index. *Sustainability, 4*, 430–461.

Barbier, E. B., Markandya, A., & Pearce, D. W. (1990). Environmental sustainability and cost-benefit analysis. *Environment and Planning A, 22*(9), 1259–1266.

Barbier, E., Koch, E. W., Silliman, B. R., Hacker, S. D., Wolanski, E., Primavera, J., Granek, E. F., Polasky, S., Aswani, S., Cramer, L. A., Stoms, D. M., Kennedy, C. J., Bael, D., Kappel, C. V., Perillo, G. M., & Reed, D. J. (2008). Coastal ecosystem-based management with nonlinear ecological functions and values. *Science, 319*, 321–323.

Bateman, I. J., Mace, G. M., Fezzi, C., Atkinson, G., & Turner, R. K. (2011). Economic analysis for ecosystem service assessments. *Environmental and Resource Economics, 48*, 177–218.

Böhnke-Henrichs, A., de Groot, R., Baulcomb, C., Hussain, S., Koss, R., & Fofana, A. (2013). Typology and indicators of ecosystem services for marine spatial planning and management. *Journal of Environmental Management, 130*, 135–145.

Borja, A., Elliott, M., Andersen, J. H., Cardoso, A. C., Carstensen, J., Ferreira, J. G., Heiskanen, A.-S., Marques, J. C., Neto, J., Teixeira, H., Uusitalo, L., Uyarra, M. C., & Zampoukas, N. (2013). Good environmental status of marine ecosystems: What is it and how do we know when we have attained it? *Marine Pollution Bulletin, 76*(1–2), 16–27.

Boyd, J., & Banzhaf, H. S. (2007). What are ecosystem services? The need for standardized environmental accounting units. *Ecological Economics, 63*, 616–626.

Chan, K. M., Kai, M. A., Satterfield, T., & Goldstien, J. (2012). Rethinking ecosystem services to better address and navigate cultural values. *Ecological Economics, 74*, 8–18.

Cooper, P. W. (2013). Socio-ecological accounting: DPSWR, a modified DPSIR framework, and its application to marine ecosystems. *Ecological Economics, 94*, 106–115.

Costanza, R., D'Arge, R., Groot, R. D., Farber, S., Grasso, M., Hannon, B., Limburg, K., Naeem, S., O'Neil, R. V., Paruelo, J., Raskin, R. G., Sutton, P., & Belt, M. V. D. (1997). The value of the world's ecosystem services and natural capital. *Nature, 387*, 253–260.

Crossland, C. J., Kremer, H. H., Lindeboom, H. J., Marshall Crossland, J. I., & Le Tissier, M. D. A. (2005). *Coastal fluxes in the anthropocene*. Berlin: Springer.

Daily, G. (1997). *Nature's services*. Covelo: Island Press.

de Groot, R. S., Fisher, B., Christie, M., Aronson, J., Braat, L., Gowdy, J., Haines-Young, R., Maltby, E., Neuville, A., Polasky, S., Portela, R., & Ring, I. (2010). Integrating the ecological and economic dimensions in biodiversity and ecosystem service valuation. In P. Kumar (Ed.), *The economics of ecosystems and biodiversity: Ecological and economic foundations* (pp. 9–40). London/Washington, DC: Earthscan.

De Jonge, V., Pinto, R., & Turner, R. K. (2012). Integrating ecological, economic and social aspects to generate useful management information under the EU Directives' ecosystem approach. *Ocean and Coastal Management, 68*, 169–188.

Elliott, M. (2011). Marine science and management means tackling exogenic unmanaged pressures and endogenic managed pressures – A numbered guide. *Marine Pollution Bulletin, 62*, 651–655.

Elliott, M. (2013). The 10-tenets for integrated, successful and sustainable marine management. *Marine Pollution Bulletin, 74*(1), 1–5.

Elliott, M., & Whitfield, A. (2011). Challenging paradigms in estuarine ecology and management. *Estuarine, Coastal and Shelf Science, 94*, 306–314.

Elliott, M., Burdon, D., Hemingway, K. L., & Apitz, S. (2007). Estuarine, coastal and marine ecosystem restoration: Confusing management and science – A revision of concepts. *Estuarine, Coastal and Shelf Science, 74*, 349–366.

Fisher, B., Turner, R. K., & Morling, P. (2009). Defining and classifying ecosystem services for decision making. *Ecological Economics, 68*, 643–653.

Fletcher, S., Saunders, J., Herbert, R., Roberts, C., & Dawson, K. (2012). *Description of the ecosystem services provided by broad-scale habitats and features of conservation importance that are likely to be protected by Marine Protected Areas in the Marine Conservation Zone Project area*. Natural England Commissioned Reports, Number 088.

Hattam, C., Atkins, J. P., Beaumont, N., Borger, T., Bohnke-Henrichs, A., Burdon, D., De Groot, R., Hoefnagel, E., Nunes, P., Piwowarczyk, J., Sergio, S., & Austen, M. (2015). Marine ecosystem services: Linking indicators to their classification. *Ecological Indicators, 49*, 61–75.

Hering, D., Borja, A., Carstensen, J., Carvalho, L., Elliott, M., Feld, C. K., Heiskanen, A. S., Johnson, R. K., Moe, J., Pont, D., Solheim, A. L., & van de Bund, W. (2010). The European water framework directive at the age of 10: A critical review of the achievements with recommendations for the future. *Science of the Total Environment, 408*, 4007–4019.

Kelble, C. R., Loomis, D. K., Lovelace, S., Nuttle, W. K., Ortner, P. B., Fletcher, P., Cook, G. S., Lorenz, J. J., & Boyer, J. N. (2013). The EBM-DPSER conceptual model: Integrating ecosystem services into the DPSIR framework. *PLoS ONE, 8*(8), e70766.

Kristrom, B. (2005). Framework for assessing the distribution of financial effects of environmental policies. In Y. Serret & N. Johnston (Eds.), *The distributional effects of environmental policy*. Cheltenham: Edward Elgar.

Krutilla, K. (2005). Using the kaldor-hicks tableau format for cost-benefit analysis. *Journal of Policy Analysis and Management, 24*, 864–875.

Lester, S. E., Costello, C., Halpern, B. S., Gaines, S. D., White, C., & Barth, J. A. (2013). Evaluating tradeoffs among ecosystem services to inform marine spatial planning. *Marine Policy, 38*, 80–89.

Mäler, K.-G., Aniyer, S., & Jansson, A. (2009). Accounting for ecosystems. *Environmental and Resource Economics, 42*, 39–51.

Mee, L. D. (2005). Assessment and monitoring requirements for the adaptive management of Europe's regional seas. In J. Vermaat, L. Bouwer, R. K. Turner, & W. Salomons (Eds.), *Managing European coasts* (pp. 227–237). Berlin: Springer.

Mee, L. D. (2012). Between the devil and the deep blue sea: The coastal zone in an era of globalisation. *Estuarine, Coastal and Shelf Science, 96*, 1–8.

Mee, L. D., Jefferson, R., Laffoley, D.'. A., & Elliott, M. (2008). How good is good? Human values and Europe's proposed marine strategy directive. *Marine Pollution Bulletin, 56*, 187–204.

Millennium Ecosystem Assessment (MEA). (2005). *Millennium ecosystem assessment: Ecosystems and human wellbeing: Synthesis*. Washington, DC: Island Press.

MMO. (2013). *Potential for co-location of activities in marine plan areas*. A report produced for the Marine Management Organisation, pp 98. MMO Project No: 1010. ISBN: 978-1-909452-08-4.

Ostrom, E. (2007). A diagnostic approach for going beyond panaceas. *Proceedings of the National Academy of Sciences, 104*, 15181–15187.

Potts, T., O'Higgins, T., Mee, L., & Pita, C. (2011). *Public perceptions of Europe's seas – A policy brief*. EU FP7 KNOWSEAS Project. ISBN 0-9529089-3-X.

Roach, B., & Wade, W. W. (2006). Policy evaluation of natural resource injuries, using habitat equivalency analysis. *Ecological Economics, 58*, 421–433.

Saunders, J., Tinch, R., & Hull, S. (2010). *Valuing the marine estate and UK seas: An ecosystem services framework*. The Crown Estate, 54 pp. ISBN: 978-1-0-906410-15-5.

Stouffer, R. J. (2012). Oceanography: Future impact of today's choices. *Nature Climate Change, 2*, 397–398.

Tett, P., Gowen, R., Painting, S., Elliott, M., Foster, R., Mills, D., Brenan, E., Capazzo, E., Fernades, T., Foden, I., Geider, R., Gilpin, L., Huxham, M., McQuatters-Gollop, A., Malcolm, S., Saux-Picart, S., Platt, T., Recalt, M. F., Suttyendranath, S., van den Moden, J., & Wilkins, M. (2013). Framework for understanding marine ecosystem health. *Marine Ecology; Progress Series, 494*, 1–27.

Turner, R. K. (1993). *Sustainable environmental economics and management: Principles and practice*. London: Belhaven press.

Turner, R. K. (1999). The place of economic values in environmental valuation. In I. J. Bateman & K. G. Willis (Eds.), *Valuing environmental preferences* (pp. 2–41). Oxford: Oxford University Press.

Turner, R. K. (2000). Integrating natural and social science in coastal management. *Journal of Marine Systems, 25*, 447–460.

Turner, R. K. (2011). *A pluralistic approach to ecosystem assessment and evaluation* [online]. A Report to Defra, London. Available from. http://www.defra.gov.uk/naturalcapitalcommittee/files/ncc-assetcheck-03.pdf

Turner, K., Lorenzoni, I., Beaumont, N., Bateman, I., Langford, I., & MacDonald, A. L. (1998). Coastal management for sustainable development: Analysing environmental and socio-economic changes on the UK coast. *The Geographical Journal, 164*, 269–281.

Turner, R. K., Paavola, J., Cooper, P., Jessamy, V., & Georgiou, S. (2003). Valuing nature: Lessons learned and future research directions. *Ecological Economics, 46*, 493–510.

UK National Ecosystem Assessment. (2011). *The UK national ecosystem assessment technical report*. Cambridge: UNEP-WCMC.
UK National Ecosystem Assessment Follow-On. (2014). *The UK national ecosystem assessment follow-on: Synthesis of the Key findings*. Cambridge, UK: UNEP-WCMC, LWEC.
UNU-IHDP & UNEP. (2012). *Inclusive wealth report 2012. Measuring progress toward sustainability*. Cambridge, UK: Cambridge University Press.
Wallace, K. J. (2007). Classification of ecosystem services: Problems and solutions. *Biological Conservation, 139*, 235–246.

# Chapter 3
# Natural Sciences Modelling in Coastal and Shelf Seas

**Tim Jickells, Julian Andrews, Steve Barnard, Paul Tett, and Sonja van Leeuwen**

## 3.1 Introduction

Effective coastal zone management requires a sound, comprehensive and integrated scientific understanding of the coastal system. Models offer a way to synthesise our understanding of the environment, analyse changes in ecosystems with complex and non-linear interactions, and forecast future changes. Such models range from conceptual frameworks, through correlation models systematically fitted to data, and rigorous and fundamental models based on physical, chemical, biological and ecological theory. Models are inevitably developed based on the conceptual understanding of the system by the community developing the models, and to the extent that such understanding is necessarily incomplete, the models will also necessarily be incomplete.

The complexity of the coastal marine environment represents an intellectual and technical challenge to observational scientists and modellers; incorporating the complexity of human interactions within that environment adds a further layer of complexity. Because of this, even the rigorous models based on fundamental

T. Jickells (✉)
Centre for Ocean and Atmospheric Sciences (COAS), School of Environmental Sciences, University of East Anglia, Norwich, UK
e-mail: t.jickells@uea.ac.uk

J. Andrews
School of Environmental Sciences, University of East Anglia, Norwich, Norfolk NR4 7TJ, UK

S. Barnard
Institute of Estuarine & Coastal Studies (IECS), University of Hull, Hull HU6 7RX, UK

P. Tett
Scottish Association for Marine Sciences (SAMS), Oban, Argyll PA37 1QA, UK

S. van Leeuwen
CEFAS, Centre for Environment, Fisheries & Aquaculture Science, Pakefield Rd, Lowestoft, NR33OHT Suffolk, UK



R.K. Turner, M. Schaafsma (eds.), *Coastal Zones Ecosystem Services*, Studies in Ecological Economics 9, DOI 10.1007/978-3-319-17214-9_3

underlying physical theory require some approximations and parameterisations, and as the range of processes and interactions increases, the number of approximations and parameterisations also increase.

It also must be recognised that models are developed for particular applications. Models can often be adapted to meet other goals, but there is no single model suitable for addressing the wide range of issues of relevance to coastal ecosystem management. Models must therefore be viewed as simplifications and abstractions of the complex environmental reality and as useful tools to help us investigate the systems and consider how they may evolve in the future, but they cannot provide a complete description of complex marine systems. For some purposes very simple models are entirely appropriate to address the research or management questions and for other purposes large complex models are required. It is not a matter that any one type of model is better than the other, but rather which is suitable for the task in question.

In the next section we offer an overview of the types of marine model tools that are likely to be useful for coastal ecosystem management. This is not intended to be a comprehensive review but does aim to show the breadth of tools that are available and necessary for this task. We will use a variety of examples of different models, and many will have been chosen to represent work we are particular familiar with focussed on the European shelf. A feature of the shelf sea environment is that it is heavily influenced by processes on and in the adjacent land and open ocean environment. This interaction can be addressed by coupling different models as discussed below, and this means that the shelf region cannot be modelled in isolation. Furthermore while the broad physical, biogeochemical and ecological principles are common to all shelf seas, the interactions of these processes with coastal geomorphology create locally distinct responses and mean that coastal models have to be geographically defined to some extent.

## 3.2 Examples of Models

We will begin with a short overview of the types of models that have been developed for various purposes in the shelf sea environment, noting their objectives and limitations, examples of their applications and anticipated future development. We then consider in the final section the extent to which these can describe the goods and services needed for effective coastal ecosystem management. In this overview we will in a broad sense move from more mathematically based physical models, through biogeochemical and ecological models, to less mathematically based decision support models that aim to synthesise a wide range of processes, pressures and attributes together to aid marine management.

### *3.2.1 Physical Ocean Models*

Physical models, based on underlying physical laws, are those used to forecast the transport of water, heat, salt and energy on a variety of time scales. Many of these are now operational models used routinely to provide forecasts and scenarios for users of the coastal and marine environment and wider society.

The first group of models considered here relate to sea-level itself and these include predictions of tidal height which have been routinely made with considerable confidence for many years. Certain meteorological conditions can lead to storm surges that lead to unusually high tides. There are operational models to predict such storm surges which have been demonstrated to have good accuracy (e.g. www.ncof.co.uk/index.htm). These models incorporate tidal predictions, detailed sea floor bathymetry and weather forecasting from the regional meteorological services to produce forecasts of actual tidal heights and warn of storm surges (Flowerdew et al. 2010), providing information for government agencies to mitigate the hazard. The success of such information is illustrated by a recent storm surge in the North Sea in December 2013, in which water levels approached (and in places exceeded) those of the infamous 1953 surge in which more than 2,500 people died. In 2013, with more accurate forecasting, extensive public warning and effective management responses, damage to property was minimised and there was no loss of life directly related to the flooding (www.metoffice.gov.uk/learning/learn-about-the-weather/weather-phenomena/storm-surge). Storm surge models are designed to deliver forecasts up to 2 days ahead, but information from these models can also be used to help predict future storm surge levels in support of the design of flood defences which have a proposed lifetime of decades (e.g. Lewis et al. 2011).

A related issue is the general rise in sea level in the future, due to global warming, which threatens coastal areas. The model projection of future global sea level rise depends on knowledge of the changes in global sea level itself, driven primarily by the warming and consequent expansion of the oceans, and the inputs of water from melting glacial ice. The effects of ocean warming on sea level can be predicted quite well, but the effects of ice sheet melt over coming decades are currently rather uncertain (IPCC 2013). Future local relative sea-level rise at any particular location also depends on the changes in the relative positions of the land surface to sealevel, which in northern (e.g. Europe, Siberia, Canada) and southern (e.g. Patagonia, Antarctica) regions of the world is primarily a response to the removal of glacial mass (isostatic adjustment) following the last glaciation. This deglaciation effect now leaves, for example, land in the north western UK rising and the south east subsiding at rates that are significant in terms of overall sea level. This land movement can be estimated from extrapolations of rates measured over recent geological past or from models of isostatic readjustment (Shennan and Woodworth 1992). The main uncertainties in these projections relate to scientific uncertainties over the impacts of climate change on ice sheet melt and uncertainties related to future emission scenarios and their impacts on climate change (Hanna et al. 2013). These sea level rise projections and their associated uncertainties can be used to estimate future flood risk by, for example, incorporation into probabilistic models (Purvis et al. 2008). Governments now publishes estimates of future sea-level rise (for instance in the UK: http://www.ukcip.org.uk/resources/ukcp09-sea-level-change) to aid planning.

A second group of physical models are the shelf sea hydrodynamic model systems which can be used to predict the circulation of heat, salt and water. Applications include estimated flows of these components within the North-West European shelf sea waters, and the exchange of these properties with offshore waters of the North

Atlantic, processes which are critical for the nutrient supply and productivity of many shelf sea systems (Huthnance 2009). Models can also provide predictions of wave height (Cavaleri et al. 2007), which is important particularly for offshore operations and which, when coupled to hydrodynamic models, can contribute to studies of sediment transport. Both types of models must be driven by meteorological models and hence can be coupled to scenarios to consider impacts of climate change.

The main UK shelf sea hydrodynamic model system has been based on the POLCOMS system for many years (Holt et al. 2009) and is now converting to the European NEMO system (www.nemo-ocean.eu) which offers some improvements over POLCOMS (O'dea et al. 2012) with a common modelling system for the ocean and coastal seas. Similar shelf sea model systems are available elsewhere, e.g. POM (www.ccpo.odu.edu/POMWEB), ROMS (www.myroms.org), HAMSOM (www.ifm.zmaw.de/research/models/hamsom), GETM (www.getm.eu) and many others. Such models are forced partly by long term averaged or modelled wind and river run off data and the quality of the output of such a model depends at least in part on the quality and accuracy of these input terms. Hydrodynamic modelling of shelf seas is challenging for many reasons, but the complex nature of the bathymetry of the shelf /ocean boundary and the associated complexity of the shelf/ocean water exchange is a particular issue (Huthnance 1995). A shelf hydrodynamic model can be coupled to other models such as marine biogeochemical models which are described below. The output from these hydrodynamic models is used to make public forecast of the impacts of climate change, for example of future changes in the temperature of the European Shelf and in the resultant hydrodynamics of the European Shelf Seas (Holt et al. 2009). Hydrodynamic models can also support marine habitat studies providing information, for instance, on bioclimatic zones and the connections between regions via ocean currents, information which is relevant to processes such as larval dispersion. For example, the recent warming of the North Sea and North Atlantic has already been linked with changes in species distribution (Beaugrand et al. 2013).

These hydrodynamic models can also be used to predict changes in water column seasonal stratification (the vertical separation of water bodies in terms of density, a separation that impedes vertical mixing) with rising temperatures, a process that is important because stratification is coupled to the development of lower oxygen conditions in near-bed waters (e.g. Diaz and Rosenberg 2008; Queste et al. 2013). Lowe et al. (2009) conclude that both the intensity and duration of such seasonal stratification is likely to increase in the future in some parts of the shelf seas, but large uncertainties in such projections arise from the assumed future greenhouse gas emission scenarios, model uncertainties and natural variability (Hawkins and Sutton 2009).

Hydrodynamic models can also be used to estimate the transport and deposition of suspended sediments within the shelf seas due to natural processes of tide and wind, and also by activities such as dredging and fishing (e.g. Luyten et al. 1999; Lee et al. 2002; van der Molen et al. 2009). These models require predictions of hydrodynamics to be coupled to descriptions of different sediment types, since the movement of different sediment types, such as coarse sand and cohesive muds, by ocean currents differ in important ways. Sediment resuspension affects many processes including the light climate and hence primary production, the transport and fate of sediment bound pollutants, sediment carbon burial, and the nature of the seabed itself, which is critically important for benthic ecology. The incorporation of sediment resuspension and deposition processes within hydrodynamic models also allows the water column transport and the development of bedforms to be predicted (Dolphin and Vincent 2009; van der Molen et al. 2004, 2009). Models are also available to describe processes of beach erosion and deposition (e.g. Bacon et al. 2007) in support of management of coastal sea defences.

### 3.2.2 *Biogeochemical and Ecological Models*

Phytoplankton (microscopic free drifting photosynthetic organisms) forms the basis of the marine food web using nutrients (e.g. nitrogen and phosphorus) and $CO_2$ to synthesise living organic matter. Models of the lower trophic levels of the marine food web require coupling of light and nutrient supply, sediment-water interactions and the ecology of phytoplankton, bacteria and zooplankton and can be linked to hydrodynamic models. These are mechanistic models (i.e. based on process understanding) driven by meteorological forcing factors, open boundary forcing factors (representing far-field influences such as exchanges with the open ocean) and nutrient forcing (from land, ocean and atmospheric sources). These models are governed by the hydrodynamic conditions, and therefore only represent the lower trophic levels (such as plankton), for which the horizontal movement of the relevant organisms is dictated by the currents. Relatively simple biogeochemical models are available which are suitable for addressing particular issues such as aquaculture (e.g. Tett et al. 2003, 2011) as well as more complex modelling systems.

One of the best developed and most extensively used of such complex biogeochemical modelling systems is the European ERSEM system (Baretta et al. 1995, Baretta 1997, Blackford et al. 2004; Edwards et al. 2012). This model describes the rates of a wide variety of processes, but is limited to a small number of general classes of ecological groups; for example four to six different broad classes of phytoplankton, based on their ecological function. The models can be run in isolation or coupled to a hydrodynamic model (e.g. van Leeuwen et al. 2013). These kinds of models can be used to consider current carbon budgets (Wakelin et al. 2012), the effects of future climate change on the lower marine trophic food web and also the impact of changes in nutrient inputs (e.g. Lenhart et al. 2010; Artioli et al. 2012; Holt et al. 2012). In an inter-comparison exercise for the North Sea, different models produced similar but not identical results, reflecting the differences in model design, and emphasising that such models are valuable, but currently limited, tools for describing the marine ecosystem and developing marine management policy (Lenhart et al. 2010). Models have been shown to be capable of simulating several ecosystem components particularly at the coarser scale (Artioli et al. 2012; Shutler et al. 2013) and the degree of uncertainty in such models can be rigorously assessed (de Mora et al. 2013).

Multi-model-ensemble approaches, in which different models are combined and compared (often resulting from international collaboration), are a key part of the IPCC strategy for assessing uncertainty in predictions of climate change (Meehl et al. 2007). This approach could usefully be widely adopted where possible for improving confidence in predicting change in marine ecosystem services where these are part of a shared marine region or subregion (e.g. Lenhart et al. 2010).

The inputs of nutrients to the coastal waters come from offshore, rivers, groundwater and the atmosphere, and the management of such inputs is an important component of marine ecosystem management. The offshore nutrient supply is currently estimated from modelled flows of water between the ocean and shelf (see above) coupled to offshore nutrient average annual concentrations based on observed (rather than modelled) distributions of nutrients. River inputs data are often based on water quality sampling and gauged river flows. These riverine chemical inputs are sampled at rather low frequency compared to their known short-term variability and in many regions sampling stations are a considerable distance inland of estuaries to avoid problems of operating gauging stations in areas of tidally reversing flow (Littlewood and Marsh 2005). This gauging station issue requires adjustments in the models to account for input and removal processes taking place below the final gauging station to provide accurate representations of inputs to the estuary itself as discussed by Jickells et al. (2014). An alternative approach to using the monitored river inputs is to model them. Sophisticated models of catchment nutrient flows are available (e.g. SWAT (Gassman et al. 2007), E-HYPE: e-hypeweb.smhi.se). However, the detailed nature of these models (making them very demanding of data and computer resources), makes it difficult to fully couple these to shelf sea models, but this is now becoming possible in systems where a few large rivers dominate (e.g. Lancelot et al. 2007). There are simpler models available to estimate global scale river nutrient fluxes, based on simplifying assumptions about inputs and generalisations about nutrient processing in catchments (e.g. Seitzinger et al. 2010). However, where directly gauged flows are available these are still preferable to model derived flows, as the latter are not yet able to simulate the full spatial and temporal variability found in nature.

Despite the uncertainties over inputs, models offer a method to evaluate the impacts of inputs that cannot be done in any other way. This is illustrated in Fig. 3.1 where the impact of particular groups of rivers on North Sea nitrogen levels, is shown using a model "experiment" in which particular rivers are "tracked" in the model so that their nutrients can be identified within the marine ecosystem, throughout the chemical and biological cycles, and the spatial effects of the rivers on the ecosystem evaluated (Lenhart et al. 2010). Such an approach allows managers to target nutrient reduction strategies more cost effectively.

The inputs of riverine and groundwater nutrients to the coastal seas are modified by estuarine processes (Statham 2012). While there is abundant evidence of modification of fluxes within estuaries, the scale and nature of these effects are poorly understood and generic models for these interactions are not available. Corrections for estuarine processes are therefore usually either based on specific models designed for a particular estuary, or on average transmission factors for each nutrient

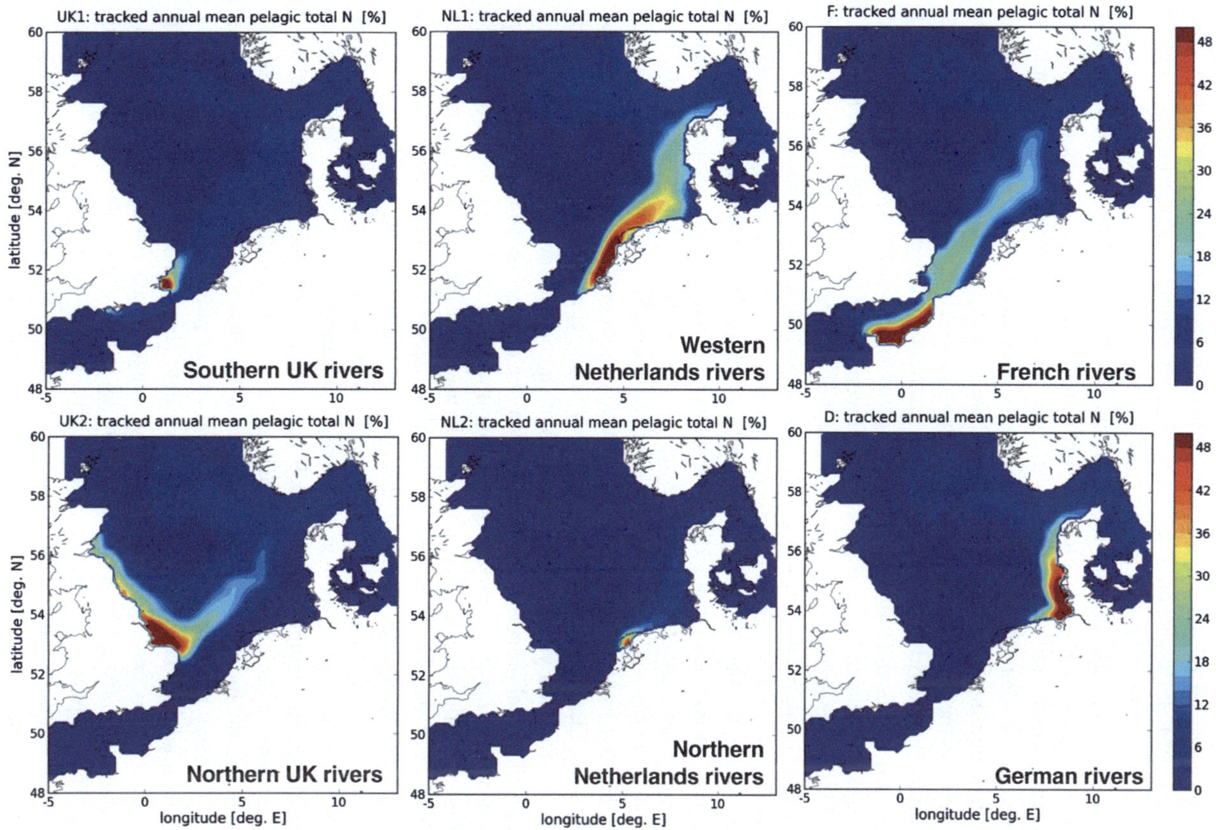


**Fig. 3.1** Contribution of different river systems to total fluvial N load to the North Sea. Unpublished cefas results (J. van der Molen) for OSPAR modelling of transboundary nutrient transport, see also OSPAR (2013)

in estuaries in general. Atmospheric inputs of the nutrient nitrogen (but not phosphorous or silicate) to coastal seas can be significant, of the order of 25 % of land based inputs for example. These atmospheric inputs can be estimated either from extrapolation of coastal data or from models of atmospheric transport and deposition such as EMEP (www.emep.int), or very rarely from direct measurements over the coastal waters (e.g. Spokes and Jickells 2005).

The distribution and cycling of contaminants in coastal seas depends on the inputs to the region, water mass transport within the region and the reactivity of the contaminant which controls its loss by degradation or transfer to the sediments or atmosphere. The distribution can therefore be described using a hydrodynamic model, provided input data are available and the reactivity of the contaminants can be described in the hydrodynamic model. Tappin et al. (2008) successfully described the distribution of various potential contaminant trace metals in the North Sea based on published inputs and a distribution coefficient to describe the partitioning of the metals between suspended sediments (which were modelled) and the water phase.

Results from hydrodynamical models coupled to biogeochemical models can in turn be linked to higher trophic level models (representing animals which control their own movement, such as fish) to assess the possible impact of pressures like climate and nutrient availability. These pressures will ultimately affect food supply and hence fish biomass or fisheries yield. Higher trophic level models can be size-structured models (e.g. Blanchard et al. 2009), species-based models (e.g. Hjollo et al. 2012) or food web models (based on multiple species characteristics and interactions, e.g. the Ecopath/Ecosim/Ecospace model suite – www.ecopath.org). These models incorporate pressures like fishing effort and fisheries management which again affect fish biomass and fisheries yield. In combination with a coupled model for hydrodynamic and biogeochemical processes, such models can be used to assess the relative impact of bottom-up and top-down pressures on economic activities and management strategies. Higher trophic level models are routinely used in fisheries management (Jordan et al. 2012) and are able to consider individual species and the impact of fisheries practice (Heymans et al. 2011) on the fish stock (Mackinson et al. 2009).

Temperature changes will also affect the distributions of fish in coastal waters, as different fish species optimise their temperature and habitat preferences. Habitat suitability and climate change impact models have therefore been developed to address this issue. Jones et al. (2012) have reviewed the outputs of three such models and concluded that the models are useful, but need to be used with reference to the model uncertainty. They therefore require additional expert judgement to deliver effective management advice. The overall effects of climate change and other environmental pressures such as ocean acidification and low oxygen levels (hypoxia) on the whole of the fish community and the commercial fishing industry have been evaluated using models. The outputs of different models give rather different results reflecting assumptions about the interactions of species but, for example, Cheung et al. (2012) used one of these model systems to suggest significant changes in fish landings by 2050 in at least one region with considerable financial implications. These type of models have also been expanded to consider interactions between species based on primary production spatial and temporal availability (Fernandes et al. 2013a).

Some models offer the opportunity to incorporate all relevant ecosystem and physical processes into the modelling system, including ecosystem services. The scale and complexity of the processes involved create major challenges for the construction and validation of such models (which are sometime called end-to-end models) and hence simplification and parameterisations are required. The key is to ensure that such simplifications are appropriate to the goals of the work. ATLANTIS is an example of such a model that is adaptable to tackling different tasks and can provide valuable information about interactions across the whole ecosystem from nutrient cycling to fish (Fulton 2010; Link et al. 2010). Heath (2012) has recently applied such a modelling approach to North Sea fisheries yields. Although this model cannot consider individual species, it does reveal the complex interplay between groups of fish (e.g. water-column feeders and bottom feeders) and other components of the environment, and provides important information to support environmental management. The complexity of the whole marine ecosystem is such that complete species level whole ecosystem models are mainly still in the development phase.

The biogeochemical models described so far are all designed to operate basically at the scale of the whole shelf sea. However, for the operation and regulation of commercial activities such as fish farms, two sorts of smaller-scale models are needed. At the fish farm scale itself, models with a high spatial resolution, for example of the order of tens of meters or better (Cromey et al. 2002; Rawson et al. 2007) are required by commercial operators and regulators. On the water-body scale, simple models such as that for 'Equilibrium Concentration Enhancement' (Gillibrand and Turrell 1997) and models of intermediate complexity (Tett et al. 2011) are useful for estimating the capacity of sealochs and estuaries to assimilate farm waste.

### 3.2.3 Bayesian Belief Networks and Decision Support Models

As the ecosystem representation and the associated models become more complex it becomes increasingly difficult to develop mechanistic models that quantitatively describe all the interactions of interest. The outputs from such large and complex model systems can also often be difficult to interpret, limiting their utility for environmental management. This has led to the development of alternative modelling and synthesis approaches that are designed to work where knowledge is incomplete and where very different sorts of information, including expert judgement as well as quantitative mechanistic or correlational relationships, need to be integrated. Examples of such approaches include Bayesian Belief Networks (probabilistic graphical models) and database or spread-sheet based integrative models.

Bayesian Belief Networks (BBNs) are models that graphically and probabilistically represent causal and statistical relationships among variables (McCann et al. 2006). BBN models are flexible integrative modelling tools, which can incorporate quantitative information that can be obtained from other models, empirical data, monitoring or specific investigations. Where data is missing, qualitative information (mostly from expert judgement) can be applied, so that the BBN becomes a flexible integrative modelling tool. The BBNs generated outputs reflect uncertainty, and can also clearly document where assumptions are made, making them a very good tool for analysis of relationships between different components and management options (Jensen and Nielsen 2007). BBNs can also deal with a wide range of problems to support decisions in environmental management, natural resources and ecosystem services, (e.g. Varis and Kuikka 1997; Marcot et al. 2001; McCann et al. 2006; Castelletti and Soncini-Sessa 2007; Henriksen et al. 2007; Uusitalo 2007; Barton et al. 2008, 2012; Fernandes et al. 2010, 2012, 2013b; Johnson et al. 2010, 2012; Haines-Young 2011; Landuyt et al. 2013), BBNs are also a valid tool for participatory environmental modelling with experts and stakeholders (Bromley et al. 2005; Henriksen et al. 2007) and can effectively integrate environmental and socio-economic considerations (Barton et al. 2012).

The word 'belief' in 'BBN model' emphasises that these models are human societal constructs. A BBN model can have two components, as exemplified in an application to the state of the north-western Black Sea as a function of land-use (in the Danube catchment) and fisheries (Langmead et al. 2009). In this study, one component was a conceptual DPSIR model (see Chap. 2) developed in expert workshops, specifying the links between key processes for which indicators were available. This can be seen as a mechanistic, albeit qualitative, model. The second component was empirical, involving a Bayesian analysis of indicator time-series that specifies probability distributions for effect variables given frequency distributions for cause variables, which were de-dimensionalised by assigning to a small number of state categories (e.g. 'low' or 'high').

Bayesian models can be based on identifying the relationships between ecosystem components as well as statistical information about those components, and then allowing the modelling software to develop the probability relationships between the components of interest. This approach has the advantage of including uncertainty estimates within the output information, as well as accommodating a lack of knowledge about particular components. However, this approach cannot model feedbacks within the system well, and the lack of mechanistic descriptions of processes means that dynamic variability in space and time within a system cannot be modelled (Langmead et al. 2009; Landuyt et al. 2013). BBN models are relatively new but are proving valuable in dealing with complex marine management issues, in particular because they (i) can incorporate expert judgement where detailed mechanistic relationships are poorly known, and (ii) because their output includes uncertainty estimates (e.g. Langmead et al. 2009). BBNs can be developed as dynamic models where temporal variability is integrated. There are also options to integrate modelling from BBNs and Geographic Information Systems (GIS) allowing spatial analysis and representation of BBN models outputs in a map (Barton et al. 2008; Li et al. 2010; Stelzenmuller et al. 2010; Johnson et al. 2012). Franzen et al. (2011) used this approach to construct a model for eutrophication in the Himmer fjord, near Stockholm. The model, of intermediate complexity, combined simple mechanistic models for estuarine exchange and nitrogen cycling with a regression model for the relationship between the concentration of total nitrogen in the fjord's water and the Secchi depth during summer. The Secchi depth, a simple measure of water clarity, increases with water transparency, and the social benefits were estimated from a study of willingness to pay for transparency, as a sign of good water quality. This approach illustrates how ecosystem services benefits might be brought into models.

Another alternative approach to synthesising multiple forms of complex information (such as a mixture of quantitative, qualitative or based on expert judgement) is via spread-sheet or data base tools and this can also be set within a spatial context using GIS. Such a system might allow, for example, the pressures and features on a particular area of seabed to be drawn together to identify if a management response is necessary and if so what that should be (see Fig. 3.2). Such an approach has the advantage of incorporating a wide variety of information of very different types into a spatially explicit format in a way that can be directly interrogated by a manager who does not have direct experience of the model development (e.g. Net Gain 2011). The system does not necessarily incorporate uncertainties (unlike BBNs), nor is it dynamic or mechanistic and it cannot explicitly include feedbacks (such as ERSEM), but can handle large and complex amounts of information within a geographic framework. BBNs and mechanistic models can be combined (Andonegi et al. 2011) and this approach has been extended to a global scale assessment of human impacts on the marine environment (Halpern et al. 2008).

Both BBNs and the spread-sheet approaches offer a very valuable way to integrate and present complex ecosystem information to support environmental management and to complement and support expert judgement, particularly where knowledge is incomplete as is almost inevitably the case when trying to look across the whole ecosystem. These approaches complement rather than replace the more mechanistic models.

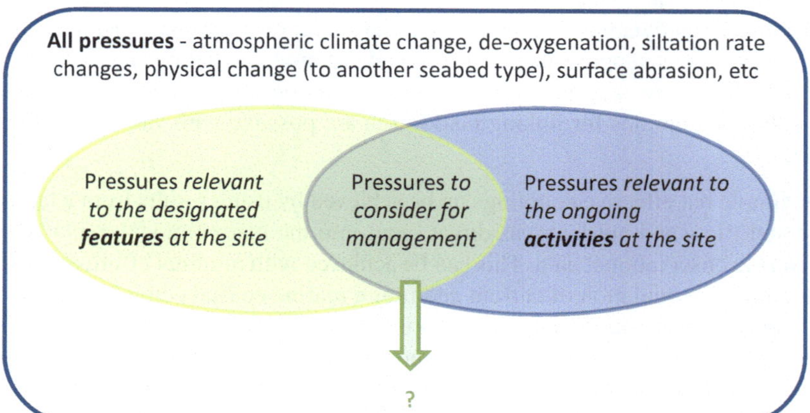


**Fig. 3.2** Illustrative example of database or spreadsheet modelling data sets that might be merged to provide management information by merging information on both the features of a site (e.g. Sediment type, biodiversity) and the activities at the site (e.g. Fisheries, sand and gravel extraction)

## 3.3 Using Models in Ecosystem Service Management

### *3.3.1 Using Models to Describe Ecosystem Services and the Goods and Benefits Derived from Them*

In Table 3.1 we tabulate selected final ecosystem services and the goods and benefits, and link them to the types of models that can contribute to the effective management of these services. The table is not designed to be an exhaustive listing of models, but rather is designed to illustrate that a wide range of models are required spanning a variety of scales and complexities to achieve this wide variety of goals, and that in some cases no models are yet available. As noted at the beginning of this section there is no single group of models suitable for this task, but rather a wide variety of tools are required in support of expert judgement and for some services, models are not really available.

### *3.3.2 Links to Land Use Change Model*

Our survey of modelling capabilities has shown that while there has been significant progress across a range of environmental contexts, the efforts so far to link terrestrial catchments to coastal and marine environments have been limited. This is clearly an important issue given that many of the drivers of change for the marine environment are terrestrially based. While estuaries are important components in such a 'coupled' approach their complexity is such that the models that do exist are site specific (Lancelot et al. 2007).

There are several options to link the terrestrial and marine environment as discussed by Torres and Uncles (2011). Until nested models are available to represent land-estuary-sea dynamics on a shelf sea scale, a simpler approach using estuarine box models may be appropriate. A box model is a model without spatial representation, which captures the main dynamics as a function of time and forcing factors. The following options for linking land to sea are possible with varying degrees of effort.

1. A simple but effective coupling can be achieved by using results from a land-use model (flow and nutrients) as direct input into the marine model, replacing the riverine observational data. This can be achieved with minimal effort, and allows for a better simulation of current and future marine coastal conditions.
2. A more comprehensive approach, taking into account estuarine processes, is to couple a land-use model to an estuarine box model based on estuarine classification (e.g. Prandle et al. 2005, 2006; Jickells et al. 2014). This approach requires some development of estuarine box models (mainly conceptual improvements), and could be included as part of the pre-processing of nutrient data before application to the marine model.

**Table 3.1** Final ecosystem services, goods and benefits and examples/types of models that can help provide information on these. Note that currently models are not really useful to assess services connected to ornamental material of genetic resources

| Final ecosystem services | Goods and benefits | Types of models |
|---|---|---|
| Fish and shellfish | Food | Wild Fisheries – fisheries yield models, biogeochemical models, end-to-end, hydrodynamic and sediment transport models, climate change models, integrative tools –BBNs and spread-sheet |
| | | Aquaculture – biogeochemical models, farm and water-body scale models |
| Algae & seaweed | Fertiliser | Macro-algae models, biogeochemical models |
| Ornamental material | Ornaments | |
| Genetic Resources | Medicines and blue technology | |
| Climate Regulation | Healthy climate | Biogeochemical models (C sequestration), climate change models, hydrodynamic models, Bayesian networks and spread-sheet |
| Natural Hazard protection | Prevention of coastal erosion and sea defences | Storm surge models, sea level rise models, sediment transport models, hydrodynamic models, BBNs and spread-sheet models |
| Clean water and sediments | Aesthetic value, food, coastal erosion | Biogeochemical models, hydrodynamic models, sediment transport models, land use models, BBNs and spread-sheet models |
| Places and seascapes | Tourism, spiritual and cultural well-being, aesthetic benefits, education | BBNs and spread-sheet models as well as more complex models can all be useful |

3. A fully coupled approach would include the estuarine box model as an extension of the marine model, allowing for both marine and land-based influences on estuarine processes. This requires development of both estuarine and marine models, and takes into account any marine representation of the estuary based on model resolution (i.e. a fine scale marine model will spatially cover more of an estuary than a coarse scale marine model, causing the estuarine box model to represent a smaller area).

Model outputs can be integrated in a BBN model to further analyse the impact of possible changes in pressures from land management (such as nutrients inputs) on ecosystem services under different scenarios. This approach can further integrate a socio-economic component and feed into a valuation assessment of ecosystem services, and hence on to the appraisal of management options as discussed in subsequent chapters.

# References


Andonegi, E., Fernandes, J. A., Quincoces, I., Irigoien, X., Uriarte, A., Perez, A., Howel, D., & Sterfanssons, G. (2011). The potential use of a gadget model to predict stock responses to climate change in combination with Bayesian networks: The case of Bay of Biscay anchovy. *ICES Journal of Marine Science, 68*, 1257–1269.

Artioli, Y., Balckford, J. C., Butenschon, M., Holt, J. T., Wakelin, S. L., Thomas, H., Borges, A. V., & Allen, J. I. (2012). The carbonate system in the North Sea: Sensitivity and model validation. *Journal of Marine Systems, 102*, 1–13.

Bacon, J. C., Vincent, C. E., Dolphin, T. J., Taylor, J. A., Pan, S. Q., & O'Connor, B. A. (2007). Shore-parallel breakwaters in meso-tidal conditions: Tidal controls on sediment transport and their longer term, regional impacts at Sea Palling, UK. *Journal of Coastal Research, 50*, 369–373.

Baretta, J. W. (1997). European regional seas ecosystem model II – Preface. *Journal of Sea Research, 38*, 169–171.

Baretta, J. W., Ebenhoh, W., & Ruardij, P. (1995). The European-regional-seas-ecosystem-model, a complex marine ecosystem model. *Netherlands Journal of Sea Research, 33*, 233–246.

Barton, D. N., Saloranta, T., Moe, S. J., Eggestad, H. O., & Kuikka, S. (2008). Bayesian belief networks as a meta-modelling tool in integrated river basin management – Pros and cons in evaluating nutrient abatement decisions under uncertainty in a Norwegian river basin. *Ecological Economics, 66*, 91–104.

Barton, D. N., Kuikka, S., Varis, O., Uusitalo, L., Henriksen, H. J., Borsuk, M., De La Hera, A., Farmani, R., Johnson, S., & Linell, J. D. (2012). Bayesian networks in environmental and resource management. *Integrated Environmental Assessment and Management, 8*, 418–429.

Beaugrand, G., McQuatters-Gollop, A., Edwards, M., & Goberville, E. (2013). Long-term responses of North Atlantic calcifying plankton to climate change. *Nature Climate Change, 3*, 263–267.

Blackford, J. C., Allen, J. I., & Gilbert, F. J. (2004). Ecosystem dynamics at six contrasting sites: A generic modelling study. *Journal of Marine Systems, 52*, 191–215.

Blanchard, J. L., Jennings, S., Law, R., Castle, M. D., McCloghrie, P., Rochet, M. J., & Benoit, E. (2009). How does abundance scale with body size in coupled size-structured food webs? *Journal of Animal Ecology, 78*, 270–280.

Bromley, J., Jackson, N. A., Clymer, O. J., Giacomello, A. M., & Jensen, F. V. (2005). The use of Hugin((R)) to develop Bayesian networks as an aid to integrated water resource planning. *Environmental Modelling & Software, 20*, 231–242.

Castelletti, A., & Soncini-Sessa, R. (2007). Bayesian networks and participatory modelling in water resource management. *Environmental Modelling & Software, 22*, 1075–1088.

Cavaleri, L., Alves, J. H. G. M., Ardhuin, F., Babanin, A., Banner, M., Belibassakis, K., Benoit, M., Donelan, M., Groeneweg, J., Herbers, T. H. C., Hwang, P., Janssen, P. A. E. M., Janssen, T., Lavrenov, I. V., Magne, R., Monbaliu, J., Onorato, M., Polnikov, V., Resio, D., Rogers, W. E., Sheremet, A., Smith, J. M., Tolman, H. L., van Vledder, G., Wolf, J., & Young, I. (2007). Wave modelling – The state of the art. *Progress in Oceanography, 75*, 603–674.

Cheung, W. W. L., Pinnegar, J., Merino, G., Jones, M. C., & Barange, M. (2012). Review of climate change impacts on marine fisheries in the UK and Ireland. *Aquatic Conservation: Marine and Freshwater Ecosystems, 22*, 368–388.

Cromey, C. J., Nickell, T. D., & Black, K. D. (2002). DEPOMOD – Modelling the deposition and biological effects of waste solids from marine cage farms. *Aquaculture, 214*, 211–239.

De Mora, L., Butenschon, M., & Allen, J. I. (2013). How should sparse marine in situ measurements be compared to a continuous model: An example. *Geoscientific Model Development, 6*, 533–548.

Diaz, R. J., & Rosenberg, R. (2008). Spreading dead zones and consequences for marine ecosystems. *Science, 321*, 926–929.

Dolphin, T., & Vincent, C. (2009). The influence of bed forms on reference concentration and suspension under waves and currents. *Continental Shelf Research, 29*, 424–432.

Edwards, K. P., Barciela, R., & Butenschom, M. (2012). Validation of the NEMO-ERSEM operational ecosystem model for the North West European continental shelf. *Ocean Science, 8*, 983–1000.

Fernandes, J. A., Irigoien, X., Goikoetxea, N., Lozano, J. A., Inza, I., Perez, A., & Bode, A. (2010). Fish recruitment prediction, using robust supervised classification methods. *Ecological Modelling, 221*, 338–352.

Fernandes, J. A., Kauppila, P., Uusitalo, L., Fleming-Lehtinen, V., Kuikka, S., & Pitkanen, H. (2012). Evaluation of reaching the targets of the water framework directive in the Gulf of Finland. *Environmental Science & Technology, 46*, 8220–8228.

Fernandes, J. A., Cheung, W. W. L., Jennings, S., Butenschon, M., De Mora, L., Frolicher, T. L., Barange, M., & Grant, A. (2013a). Modelling the effects of climate change on the distribution and production of marine fishes: Accounting for trophic interactions in a dynamic bioclimate envelope model. *Global Change Biology, 19*, 2596–2607.

Fernandes, J. A., Lozano, J. A., Inza, I., Irigoien, X., Perez, A., & Rodriguez, J. D. (2013b). Supervised pre-processing approaches in multiple class variables classification for fish recruitment forecasting. *Environmental Modelling & Software, 40*, 245–254.

Flowerdew, J., Horsburgh, K., Wilson, C., & Mylne, K. (2010). Development and evaluation of an ensemble forecasting system for coastal storm surges. *Quarterly Journal of the Royal Meteorological Society, 136*, 1444–1456.

Franzen, F., Kinell, G., Walve, J., Elmgren, R., & Soderqvist, T. (2011). Participatory social-ecological modeling in eutrophication management: The case of Himmerfjarden, Sweden. *Ecology and Society, 16*, 27.

Fulton, E. A. (2010). Approaches to end-to-end ecosystem models. *Journal of Marine Systems, 81*, 171–183.

Gassman, P. W., Reyes, M. R., Green, C. H., & Arnold, J. G. (2007). The soil and water assessment tool: Historical development, applications, and future research directions. *Transactions of the ASABE, 50*, 1211–1250.

Gillibrand, P. A., & Turrell, W. R. (1997). The use of simple models in the regulation of the impact of fish farms on water quality in Scottish sea lochs. *Aquaculture, 159*, 33–46.

Haines-Young, R. (2011). Exploring ecosystem service issues across diverse knowledge domains using Bayesian belief networks. *Progress in Physical Geography, 35*, 681–699.

Halpern, B. S., Walbridge, S., Selkoe, K. A., Kappel, C. V., Micheli, F., D'Agrosa, C., Bruno, J. F., Casey, K. S., Ebert, C., Fox, H. E., Fujita, R., Heinemann, D., Lenihan, H. S., Madin, E. M. P., Perry, M. T., Selig, E. R., Spalding, M., Steneck, R., & Watson, R. (2008). A global map of human impact on marine ecosystems. *Science, 319*(5865), 948–952.

Hanna, E., Navarro, F. J., Pattyn, F., Domingues, C. M., Fettweis, X., Ivins, E. R., Nicholls, R. J., Ritz, C., Smith, B., Tulaczyk, S., Whitehouse, P. L., & Zwally, H. J. (2013). Ice-sheet mass balance and climate change. *Nature, 498*, 51–59.

Hawkins, E., & Sutton, R. (2009). The potential to narrow uncertainty in regional climate predictions. *Bulletin of the American Meteorological Society, 90*, 1095.

Heath, M. R. (2012). Ecosystem limits to food web fluxes and fisheries yields in the North Sea simulated with an end-to-end food web model. *Progress in Oceanography, 102*, 42–66.

Henriksen, H. J., Rasmussen, P., Brandt, G., Von Bulow, D., & Jensen, F. V. (2007). Public participation modelling using Bayesian networks in management of groundwater contamination. *Environmental Modelling & Software, 22*, 1101–1113.

Heymans, J. J., Mackinson, S., Sumaila, U. R., Dyck, A., & Little, A. (2011). The impact of subsidies on the ecological sustainability and future profits from North Sea fisheries. *PLoS One, 6*, e20239.

Hjollo, S. S., Huse, G., Skogen, M. D., & Melle, W. (2012). Modelling secondary production in the Norwegian Sea with a fully coupled physical/primary production/individual-based *Calanus finmarchicus* model system. *Marine Biology Research, 8*, 508–526.

Holt, J., Harle, J., Proctor, R., Michel, S., Ashworth, M., Batstone, C., Allen, I., Holmes, R., Smyth, T., Haines, K., Bretherton, D., & Smith, G. (2009). Modelling the global coastal ocean. *Philosophical Transactions of the Royal Society a-Mathematical Physical and Engineering Sciences, 367*, 939–951.

Holt, J., Butenschon, M., Wakelin, S. L., Artioli, Y., & Allen, J. I. (2012). Oceanic controls on the primary production of the northwest European continental shelf: Model experiments under recent past conditions and a potential future scenario. *Biogeosciences, 9*, 97–117.

Huthnance, J. M. (1995). Circulation, exchange and water masses at the ocean margin: The role of physical processes at the shelf edge. *Progress in Oceanography, 35*, 353–431.

Huthnance, J. M. (2009). The northeast Atlantic Margins. In K. K. Liu, L. Atkinson, R. Quinones, L. Talaue-McManus (Eds.), *Carbon and nutrient fluxes in continental margins: A global synthesis*. Berlin: Springer.

International Multiple Sclerosis Genetics Consortium, & Wellcome Trust Case Control Consortium 2. (2011). Genetic risk and a primary role for cell-mediated immune mechanisms in multiple sclerosis. *Nature, 476*(7359), 214–219.

IPCC. (2013). *Climate change 2013: The physical science basis. Contribution of working group 1 to the fifth assessment report of the intergovernmental panel on climate change*. Cambridge: Cambridge University Press.

Jensen, F. V., & Nielsen, T. D. (2007). *Bayesian networks and decision graphs*. New York: Springer.

Jickells, T. D., Andrews, J. E., Parkes, D. J., Suratman, S., Aziz, A. A., & Hee, Y. Y. (2014). Nutrient transport through estuaries: The importance of geography. *Estuarine and Coastal Shelf Science, 150*, 215–229.

Johnson, S., Fielding, F., Hamilton, G., & Mengersen, K. (2010). An integrated Bayesian network approach to lyngbya majuscula bloom initiation. *Marine Environmental Research, 69*, 27–37.

Johnson, S., Low-Choy, S., & Mengersen, K. (2012). Integrated Bayesian networks and geographic information systems: Good practice examples. *Integrated Environmental Assessment and Management, 8*, 473–479.

Jones, M. C., Dye, S. R., Pinnegar, J. K., Warren, R., & Cheung, W. W. L. (2012). Modelling commercial fish distributions: Prediction and assessment using different approaches. *Ecological Modelling, 225*, 133–145.

Jordan, S. J., O'Higgins, T., & Dittmar, J. A. (2012). Ecosystem services of coastal habitats and fisheries: Multiscale ecological and economic models in support of ecosystem-based management. *Marine and Coastal Fisheries, 4*, 573–586.

Lancelot, C., Gypens, N., Billen, G., Garnier, J., & Roubeix, V. (2007). Testing an integrated river-ocean mathematical tool for linking marine eutrophication to land use: The phaeocystis-dominated Belgian coastal zone (Southern North Sea) over the past 50 years. *Journal of Marine Systems, 64*, 216–228.

Landuyt, D., Broekx, S., D'Hondt, R., Engelen, G., Aertsens, J., & Goethals, P. L. M. (2013). A review of Bayesian belief networks in ecosystem service modelling. *Environmental Modelling & Software, 46*, 1–11.

Langmead, O., McQuattars-Gollop, A., Mee, L. D., Friedrich, J., Gilbert, A. J., Gomoiu, M. T., Jackson, E. L., Knudsen, S., Minicheva, G., & Todorova, V. (2009). Recovery or decline of the northwestern Black Sea: A societal choice revealed by socio-ecological modelling. *Ecological Modelling, 220*, 2927–2939.

Lee, J. Y., Tett, P., Jones, K., Luyten, P., Smith, C., & Wild-Allen, K. (2002). The PROWQM physical-biological model with benthic-pelagic coupling applied to the northern North Sea. *Journal of Sea Research, 48*, 287–331.

Lenhart, H. J., Mills, D. K., Baretta-Bekker, H., Van Leeuwen, S. M., van der Molen, J., Baretta, J. W., Blaas, M., Desmit, X., Kuhn, W., Lacroix, G., Los, H. J., Menesguen, A., Neves, R., Proctor, R., Ruardij, P., Skogen, M. D., Vanhoutte-Brunier, A., Villars, M. T., & Wakelin, S. L. (2010). Predicting the consequences of nutrient reduction on the eutrophication status of the North Sea. *Journal of Marine Systems, 81*, 148–170.

Lewis, M., Horsburgh, K., Bates, P., & Smith, R. (2011). Quantifying the uncertainty in future coastal flood risk estimates for the UK. *Journal of Coastal Research, 27*, 870–881.

Li, L. F., Wang, J. F., Leung, H., & Jiang, C. S. (2010). Assessment of catastrophic risk using Bayesian network constructed from domain knowledge and spatial data. *Risk Analysis, 30*, 1157–1175.

Link, J. S., Fulton, E. A., & Gamble, R. J. (2010). The northeast US application of ATLANTIS: A full system model exploring marine ecosystem dynamics in a living marine resource management context. *Progress in Oceanography, 87*, 214–234.

Littlewood, I. G., & Marsh, T. J. (2005). Annual freshwater river mass loads from Great Britain, 1975–1994: Estimation algorithm, database and monitoring network issues. *Journal of Hydrology, 304*, 221–237.

Lowe, J., Howard, T., Pardaens, A., Tinker, J., Holt, J., Wakeling, Milne, G., Leake, J., Wolk, J., Horsburgh, K., Reeder, R., Jenkins, G., Ridley, J., Dye, S., & Bradley, S. (2009). *UK Climate projections science report: Marine and coastal projections*. UKCP09 Scientific Report.

Luyten, P. J., Jones, J. E., Proctor, R., Tabor, A., Tett, P., & Wild-Allen, K. (1999). *COHERENS- a coupled hydrodynamical-ecological model fro regional and shelf seas: User documentation*. Management Unit of the Mathematical Models of the North Sea Report, Brussels.

Mackinson, S., Daskalov, G., Heymans, J. J., Neira, S., Arancibia, H., Zetina-Rejon, M., Jiang, H., Cheng, H. Q., Coll, M., Arreguin-Sanchez, F., Keeble, K., & Shannon, L. (2009). Which forcing factors fit? Using ecosystem models to investigate the relative influence of fishing and changes in primary productivity on the dynamics of marine ecosystems. *Ecological Modelling, 220*, 2972–2987.

Marcot, B. G., Holthausen, R. S., Raphael, M. G., Rowland, M. M., & Wisdom, M. J. (2001). Using Bayesian belief networks to evaluate fish and wildlife population viability under land management alternatives from an environmental impact statement. *Forest Ecology and Management, 153*, 29–42.

McCann, R. K., Marcot, B. G., & Ellis, R. (2006). Bayesian belief networks: Applications in ecology and natural resource management. *Canadian Journal of Forest Research-Revue Canadienne De Recherche Forestiere, 36*, 3053–3062.

Meehl, G. A., Stocker, T. F., Collins, W. D., Friedlingstein, P., Gaye, A. T., Gregory, J. M., Kitoh, A., Knutti, R., Murphy, J. M., Noda, A., Raper, S. C. B., Watterson, I. G., Weaver, A. J., & Zhao, Z.-C. (2007). Global Climate Projections. In S. Solomon, D. Qin, M. Manning, Z. Chen, M. Marquis, K. B. Averyt, M. Tignor, & H. L. Miller (Eds.), *Climate change 2007: The physical science basis. Contribution of working group I to the fourth assessment report of the intergovernmental panel on climate change*. Cambridge/New York: Cambridge University Press.

Net Gain. (2011). *The North Sea marine conservation zones project; final recommendations, submission to natural England & JNCC*, 31 Aug 2011. http://jncc.defra.gov.uk/PDF/120718_MCZAP_120702_NG_Final_report_version_1.2.pdf

O'dea, E. J., Arnold, A. K., Edwards, K. P., Fumer, R., Hyder, P., Martin, M. J., Siddorn, J. R., Storkey, D., While, J., Holt, J. T., & Liu, H. (2012). An operational ocean forecast system incorporating NEMO and SST data assimilation for the tidally driven European North-West shelf. *Journal of Operational Oceanography, 5*, 3–17.

OSPAR. (2013). *Distance to target modelling assessment*. OSPAR report 599, 978-1-909159-32-7. www.ospar.org

Prandle, D., Lane, A., & Manning, A. J. (2005). Estuaries are not so unique. *Geophysical Research Letters, 32*, L23614, 1–5.

Prandle, D., Lane, A., & Manning, A. J. (2006). New typologies for estuarine morphology. *Geomorphology, 81*, 309–315.

Purvis, M. J., Bates, P. D., & Hayes, C. M. (2008). A probabilistic methodology to estimate future coastal flood risk due to sea level rise. *Coastal Engineering, 55*, 1062–1073.

Queste, B. Y., Fernand, L., Jickells, T. D., & Heywood, K. J. (2013). Spatial extent and historical context of North Sea oxygen depletion in August 2010. *Biogeochemistry, 113*, 53–68.

Rawson, M., Chen, C., Wang, D., Yarish, C., Sullivan, J., Wang, L., & Zhu, M. (2007). Ecosystem based management and models in sustainable management of coastal aquaculture. *Bulletin of Fisheries Research Agency, 19*, 97–111.

Seitzinger, S. P., Mayorga, E., Bouwman, A. F., Kroeze, C., Beusen, A. H. W., Billen, G., Van Drecht, G., Dumont, E., Fekete, B. M., Garnier, J., & Harrison, J. A. (2010). Global river nutrient export: A scenario analysis of past and future trends. *Global Biogeochemical Cycles, 24*, GB0A08.

Shennan, I., & Woodworth, P. L. (1992). A comparison of late Holocene and 20th-century Sea-level trends from the UK and North-Sea region. *Geophysical Journal International, 109*, 96–105.

Shutler, J. D., Land, P. E., Brown, C. W., Findlay, H. S., Donlon, C. J., Medland, M., Snooke, R., & Blackford, J. C. (2013). Coccolithophore surface distributions in the North Atlantic and their modulation of the air-sea flux of $CO_2$ from 10 years of satellite earth observation data. *Biogeosciences, 10*, 2699–2709.

Spokes, L. J., & Jickells, T. D. (2005). Is the atmosphere really an important source of reactive nitrogen to coastal waters? *Continental Shelf Research, 25*, 2022–2035.

Statham, P. J. (2012). Nutrients in estuaries – An overview and the potential impacts of climate change. *Science of the Total Environment, 434*, 213–227.

Stelzenmuller, V., Lee, J., Garnacho, E., & Rogers, S. I. (2010). Assessment of a Bayesian belief network-GIS framework as a practical tool to support marine planning. *Marine Pollution Bulletin, 60*, 1743–1754.

Tappin, A. D., Statham, P. J., Burton, J. D., & Gellers-Barkmann, S. (2008). Distributions and fluxes of contaminant metals in the North Sea: Comparisons between field measurements and model simulations using NOSTRADAMUS. *Environmental Chemistry, 5*, 51–67.

Tett, P., Gilpin, L., Svendsen, H., Erlandsson, C. P., Larsson, U., Kratzer, S., Fouilland, E., Janzen, C., Lee, J. Y., Grenz, C., Newton, A., Ferreira, J. G., Fernandes, T., & Scory, S. (2003). Eutrophication and some European waters of restricted exchange. *Continental Shelf Research, 23*, 1635–1671.

Tett, P., Portilla, E., Gillibrand, P. A., & Inall, M. (2011). Carrying and assimilative capacities: The ACExR-LESV model for sea-loch aquaculture. *Aquaculture Research, 42*, 51–67.

Torres, R., & Uncles, R. (2011). Modelling of estuarine and coastal waters. In E. Wolanski & D. S. McLusky (Eds.), *Treatise on Estuarine and Coastal Science* (Vol. 2, pp. 395–427). Waltham: Academic Press.

Uusitalo, L. (2007). Advantages and challenges of Bayesian networks in environmental modelling. *Ecological Modelling, 203*, 312–318.

van der Molen, J., Gerrits, J., & De Swart, H. E. (2004). Modelling the morphodynamics of a tidal shelf sea. *Continental Shelf Research, 24*, 483–507.

van der Molen, J., Bolding, K., Greenwood, N., & Mills, D. K. (2009). A 1-D vertical multiple grain size model of suspended particulate matter in combined currents and waves in shelf seas. *Journal of Geophysical Research: Earth Surface, 114*, F01030.

Van Leeuwen, S. M., van der Molen, J., Ruardij, P., Fernand, L., & Jickells, T. (2013). Modelling the contribution of deep chlorophyll maxima to annual primary production in the North Sea. *Biogeochemistry, 113*, 137–152.

Varis, O., & Kuikka, S. (1997). BeNe-EIA: A Bayesian approach to expert judgment elicitation with case studies on climate change impacts on surface waters. *Climatic Change, 37*, 539–563.

Wakelin, S. L., Holt, J. T., Blackford, J. C., Allen, J. I., Butenschon, M., & Artioli, Y. (2012). Modeling the carbon fluxes of the northwest European continental shelf: Validation and budgets. *Journal of Geophysical Research-Oceans, 117*, C05020.

# Chapter 4
# Valuation of Ecosystem Services

**R.K. Turner and M. Schaafsma**

## 4.1 Introduction: Monetary Valuation and Its Critics

Understanding the economic value of nature and the services it provides to society is important for local, national and global policy and decision making. But economic methods and tools have to be accompanied by methods and tools from a range of social and natural science disciplines if a robust DSS is to be constructed. The position taken in this book is that the core decision criterion in conventional economics, i.e. economic efficiency, is important, especially as environmental goods and services scarcity increase over time. Cost benefit analysis can, we would argue, still play an important role in multi-criteria assessment DSSs if suitably adjusted for equity concerns. Nevertheless, economic efficiency is not a meta-ethical criterion. The full commodification of all ecosystem services through the assignment of monetary values to all aspects of ecosystem complexity is not meaningful or possible and does not provide a sound scientific or moral basis for sustainable management.

Critics of the ecosystem services approach have warned against the use of the approach as an over-arching framework for policy if it is applied without recognising ecosystem complexity, scientific uncertainty and the existence of environmental limits and threshold effects (Norgaard 2010). Some go further and argue that the history of conceptual and methodological problems with monetary valuation of the environment going back to the 1960s indicates that fundamental problems are

R.K. Turner (✉)
School of Environmental Sciences, University of East Anglia, Norwich, UK
e-mail: r.k.turner@uea.ac.uk

M. Schaafsma
Geography and Environment and Centre for Biological Sciences, University of Southampton, University Road, Southampton SO17 1BJ, UK



R.K. Turner, M. Schaafsma (eds.), *Coastal Zones Ecosystem Services*,
Studies in Ecological Economics 9, DOI 10.1007/978-3-319-17214-9_4

unlikely to be resolved and that ecosystem services thinking should be abandoned (Baveye et al. 2013).

The DSS (and the Balance Sheets Approach) set out in this book for coastal management does include due recognition of system complexity, uncertainty and threshold effect risks, together with regard for limits and critical natural capital stocks. It further accepts that some ecosystem services such as some cultural services are not best expressed in monetary terms and that the DSS should be a multi-criteria process. But within these boundaries we argue that in real world policy contexts trade-offs are continually made between conservation and development options and monetary and opportunity cost calculations can and do play a useful role.

## 4.2 Values in Nature

In the economic literature, a number of issues can be identified as key to the *appropriate* economic valuation of ecosystem services. These are: spatial and policy context explicitness, marginality, the double-counting trap, non-linearities in benefits, and threshold effects (see Fig. 4.1).

Therefore to be most useful for policy, services must be assessed within their appropriate spatial and policy context and economic valuation should provide marginal estimates of value (avoiding double counting) that can feed into decisions at the appropriate scale, and which recognise possible non-linearities and are well within the bounds of safe minimum standards (MEA 2005; Turner et al. 2003).

Some ecologists use the term *value* to mean 'that which is desirable or worthy of esteem for its own sake; something or some quality having *intrinsic* worth'. Some economists use the same term to describe 'a fair or proper equivalent in money, commodities, etc', where *equivalent in money* represents that sum of money that would have an equivalent effect on the welfare or utilities of individuals. A number of ecosystem goods can be valued in economic terms, while others cannot because of uncertainty and complexity conditions. The notion of total economic value (TEV) provides an all-encompassing measure of the *economic value* of any environmental asset. It is important to note however that TEV is always less than total systems value (TSV). A minimum configuration of ecosystem structure and process is required before final services and goods can be provided and some values cannot be expressed in economic terms. We take a closer look at the TEV concept and related issues in next.

Total Economic Value (TEV) decomposes into use and non-use (or passive use) values but it does not encompass other kinds of values, such as intrinsic values which are usually defined as values residing 'in' the asset and unrelated to human preferences or even human observation (Turner 1999). However, apart from the problems of making the notion of intrinsic value operational, it can be argued that some people's willingness to pay (WTP) for the conservation of an asset, independently of any use they make of it, is influenced by their own judgements about intrinsic value (Morse-Jones et al. 2012). This may show up especially in notions of 'rights to existence' but also as a form of altruism.

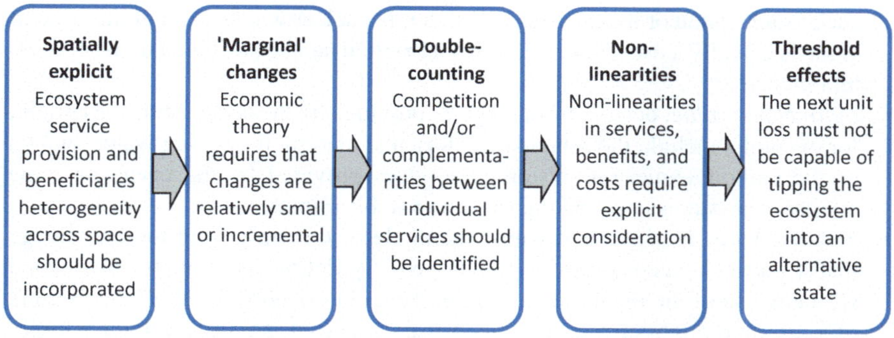


**Fig. 4.1** Ecosystem services sequential steps: a framework for appropriate economic valuation (Adapted from Turner et al. (2010))

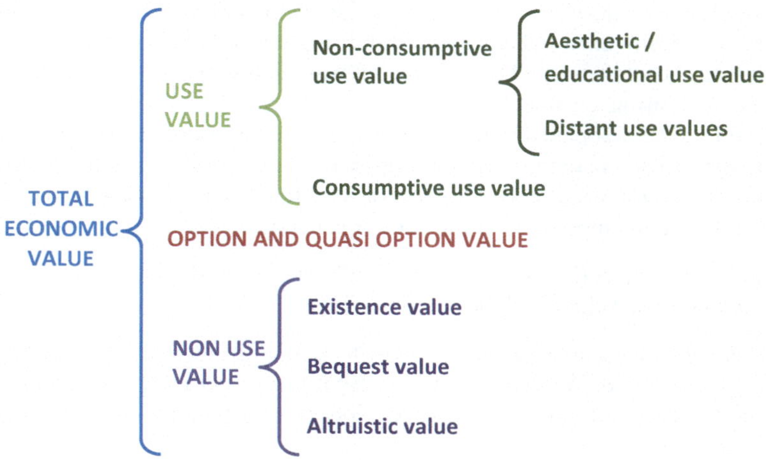


**Fig. 4.2** Total economic value

As Chap. 2 illustrated, coastal and marine ecosystems provide a wide range of final services and related benefits of significant value to society. The use of the TEV classification enables the values to be usefully broken down into the categories shown in Fig. 4.2. The initial distinction is between individual *use value* and *non-use value*. Use value involves some interaction with the resource, either directly or indirectly:

- *Direct use value*: involves direct interaction with the ecosystem itself rather than via the services it provides. It may be consumptive use, such as fisheries, or it may be non-consumptive, as with some recreational and educational activities. There is also the possibility of deriving value from 'distant use' through media such as television or magazines, although it is unclear whether or not this type of value is actually a use value, and to what extent it can be attributed to the ecosystem involved;

- *Indirect use value*: derives from services provided by the ecosystem. This might, for example, include the removal of nutrients, thereby improving water quality, or the carbon sequestration/storage services provided by the ocean or some coastal ecosystems contributing to a more stable climate;
- *Non-use value* is associated with benefits derived simply from the knowledge that a particular ecosystem is maintained. By definition, it is not associated with any use of the resource or tangible benefit derived from it, although users of a resource might also attribute non-use value to it. Non-use value is closely linked to ethical concerns, often being linked to altruistic preferences, although according to some analysts it stems ultimately from self-interest. It can be split into three basic components, which may overlap depending upon exact definitions;
- *Existence value*: derived simply from the satisfaction of knowing that an ecosystem continues to exist, whether or not this might also benefit others. This value notion has been interpreted in a number of ways and seems to straddle the instrumental and intrinsic value divide;
- *Bequest value*: associated with the knowledge that a resource will be passed on to descendants to maintain the opportunity for them to enjoy it in the future; and
- *Altruistic value*: associated with the satisfaction from ensuring resources are available to contemporaries of the current generation.

Finally, two categories not associated with the initial distinction between use values and non-use values include:

- *Option value*: an individual derives benefit from ensuring that a resource will be available for *use in the future*. In this sense it is a form of use value, although it can be regarded as a form of insurance to provide for possible future but not current use; and
- *Quasi-option value (QOV):* associated with the potential benefits of waiting for improved information before giving up the option to preserve a resource for future use. In particular, it suggests a value of avoiding irreversible damage that might prove to have been unwarranted in the light of further information. An example of an option value is in bio-prospecting, where biodiversity may be maintained on the off-chance that it might in the future be the source of important new medicinal drugs. Potentially, QOV could make up a sizeable proportion of TEV, although measurement of its magnitude is problematic.

These various elements of total economic value are assessed using economic valuation methods, and some of these elements are more easily valued than others, especially those with easily identifiable uses (usually the use type values). Non-use values are usually more difficult to assess.

## 4.3 Financial Versus Economic Values

In any socio-economic assessment it is necessary to distinguish between financial accounting and economic values and analysis. Prices and values are not necessarily equivalent and price is only that portion of the underlying value of a good which is realised in the market place (Pearce and Turner 1990). For those goods produced and consumed under reasonably competitive market conditions, their market prices are an acceptable approximation for their value, provided that there are no other prevailing distortions such as government tax or subsidy interventions. Prices will typically diverge from values when so-called public goods (with non-exclusion and non-rivalry in consumption characteristics) are involved which lack private ownership; or when the full costs of production and consumption (especially environmental impact costs) are not readily included in the pricing process. For many ecosystem (service-related) goods there are no markets available, or the full cost of their supply are not reflected in financial measures. Economic analysis seeks to uncover the value in monetary terms (and ultimately the economic welfare effect on humans) of the good in question rather than just its financial price. It measures value (welfare) through an approximation known as WTP for changes in the provision of the good. Note that this WTP measure is not the same thing as actual payment (market price); when the latter is less than the former a consumer gains value (consumer surplus).

## 4.4 Stock Versus Flow Values

The distinction between ecosystem services stocks and flows has also to be reflected in the economic valuation approach adopted. The paper in the journal Nature by Costanza et al. (1997) estimated the value of global ecosystem services at USD 33 trillion and led to a protracted debate and controversy over the 'true' value of the natural environment.

The Costanza et al. (1997) global ecosystem services estimation has been attacked on a number of grounds including that the aggregate value was not necessarily the sum of the parts, and that USD 33 trillion was more than global income and therefore peoples' ability to pay (Heal 2000). Further work (Howarth and Farber 2002) sought to defend the Costanza et al. approach by arguing that the estimates of ecosystem services value were analogous to National Income Accounting entities such as GDP with a constant set of value weights. The underlying rationale here is that the aggregate measure is a quantity parameter (the stock concept), and, while it is related to value, it does not directly value the planet's ecosystem services in total. In this sense it is an accounting price measure of the quantity of ecosystem services holding prices constant, where the measures are not based on economic theory but on accounting rules (Costanza et al. 2014). In this stock accounting context the criticism related to peoples' budget constraint and ability to pay is not relevant, because the measure is based on virtual (not real) prices and virtual incomes (i.e. incomes adjusted to enable individuals to hypothetically pay for the services).

For the income and expenditure accounts to balance, the total expenditure must be less than actual and virtual income. The current extent of European coastal blue carbon (the carbon storage service provided by salt marshes and sea grasses) has, for example, an accounting stock price (value) of about USD 180 million, valued against a Social Cost of Carbon estimate (Luisetti et al. 2013). Such total (stock) values can be estimated and compared for two different points in time as a heuristic to help to appreciate the change in natural capital. This viewpoint is, however, controversial and is not supported by many mainstream economists. For them the only relevant measure is the marginal economic value.

For economic valuation (as opposed to accounting) it is important to be able to quantify and evaluate gains or losses in stock assets and consequent service flows (analogous to net GDP).[1] Now instead of holding prices constant, we seek to determine marginal economic value as it relates to an incremental increase in a set of ecosystem services over time and space. When the ecosystem final services value relates only to non-market services, it can be combined with GDP (in the same way as relevant pollution and other externalities are internalised) to yield a more green GDP measure. However, to avoid double counting, when ecosystem goods are (partly) marketed, the value of these flows is already (partly) captured in GDP. The present value of a *discounted* flow of ecosystem services values can contribute to stock of wealth accounts, such as the Inclusive Wealth account (UNU-HDP and UNEP 2012). An important consideration is that the flow and stock values (i.e. the accounting and economic values) as explained above serve different purposes, and they are not comparable and should not be added up.

The ecosystem services valuation studies reviewed in Chap. 6 all provide estimates of the economic value per year, i.e. flow values. But a separate and complementary ecosystem services account or index may also be a worthwhile objective. Overall, the future goal should be to measure and value both service flows and to predict changes in stocks (ecosystem health) which condition future flows.

## 4.5 Monetary Valuation Techniques

A number of valuation techniques have been developed to estimate individual monetary (economic) values of flows of ecosystem goods and services. They range from (adjusted) market prices, through productivity effect (or production function) methods and revealed preference (based on consumer actions) to survey-based expressed preference methods.

Different economic valuation techniques will be appropriate for different ecosystem goods and benefits, but it will not be possible to place meaningful monetary values on all the benefits (and some of the costs) of outputs from the coastal and

[1] GDP reflects the financial (market) value of all final goods and services produced within a country within a certain period. Net Domestic Product (net GDP) is GDP net of the depreciation on capital goods, and thereby reflects how much capital has been consumed over the year.

marine zone. In particular the symbolic and cultural values assigned to some coastal and marine features and land/seascapes lie outside the monetary calculus and are conditioned by social preferences and norms (shared values) arrived at, over time, through various forms of information transmission, art, literature etc. (see Sect. 4.6). Table 4.1 gives an overview of the different individual monetary valuation methods that can be used for valuation of coastal and marine ecosystem goods and benefits, including the human welfare measure they are based on, and the goods and benefits they can be applied to.

In appraisal techniques, such as CBA and MCA, where the societal benefits of different ecosystem services and goods need to be compared, it is often necessary to use different (existing) primary valuation studies. One important issue to consider in such cases is that studies may use different valuation methods. Valuation methods do not necessarily address similar constructs of welfare, for example, producer surplus versus consumer surplus, or net versus gross revenues. Different welfare constructs, strictly speaking, cannot be added up or compared – they are different types of estimates. Another word of caution concerns the use of cost-based approaches. The replacement cost approach looks at the costs of replacing an ecosystem service by a manmade alternative (either a technology or re-created habitat). This approach assumes that if society is willing to pay these replacement costs, the value of the ecosystem benefit must be *at least* that amount, and they may be higher. As such, these cost-based estimates provide a lower bound estimate of the societal value of ecosystem goods and benefits. Cost-based methods are commonly used for the valuation of carbon sequestration benefits (healthy climate), e.g. based on the abatement cost method (looking at the cost of measures to reduce emissions), or the costs of damages avoided reflected in Social Cost of Carbon (SCC) estimates (that capture welfare changes associated with the impact of climate change), which are both typically higher than carbon market prices.

In order to scope the uncertain future outcomes, scenario analysis is often deployed in which a change from the baseline to a future state of the world is considered, and marginal economic values can be used to assign monetary values to these changes and assess the changes in welfare over time. Marginal economic values, relating to an incremental change in ecosystem service provision, are grounded in economic theory. By their nature, valuation methods differ in terms of the unit used to represent value estimates: some methods result in a value per unit area or physical (qualitative or quantitative) change in ecosystem service delivery, other studies will provide a value per household or individual for a (small set of) change(s) in area or ecosystem services provision. Some studies provide total values for an entire habitat area, for instance, the total value of fisheries along the UK coast. For scenario analyses working with land use change maps, marginal values per unit of area are most practical. However, marginal values may not be proportional to biophysical unit changes (quantity, quality, area). There is no a priori expectation about the relationship between ecosystem change and welfare change. For incremental changes in ecosystem service provision, prices (values) are expected to increase as supply decreases, but some ecosystems will have thresholds below which no services are provided. Marginal economic values are not applicable when the ecosystem

**Table 4.1** Overview of environmental valuation methods for estimating individual values (Based on Brander et al. (2006) and Turner et al. (2010))

| Method | Short description | Limitations | Welfare measure | Good/benefit |
|---|---|---|---|---|
| Travel cost method | Recreational benefits. Indirect method. Estimate demand (WTP) using travel costs to visit site. | Large data requirements, complex when trips are multipurpose, only for use values | Consumer surplus | Recreational angling, beach visits, diving, other recreational activities |
| Hedonic pricing method | Amenity benefits. Indirect method. Estimate WTP using price differentials and characteristics of related products. | Large data requirements, sensitive to model specification, only for use values | Consumer surplus | Amenity – property (housing, hotels, land) |
| Contingent valuation | All goods, also non-use values. Direct survey-based method. Hypothetical questions to obtain WTP. | Time and cost intensive, biases related to non-compensatory behaviour, constructed preferences and framing effects. | Compensating or equivalent surplus | Appreciation of culture, heritage, recreation, landscape, biodiversity. |
| | | | | Bundle of services |
| Choice experiments | All goods, direct method, also non-use values. Hypothetical questions to obtain WTP | As for CVM, and: greater cognitive burden, and associated learning and fatigue biases. | Compensating or equivalent surplus | Appreciation of culture, heritage, recreation, landscape, biodiversity. |
| | | | | Bundle of services |
| Net factor income | Assign value as revenue of an associated product net of costs of other inputs | Only applicable to marketed goods, tends to overestimate values. | Producer surplus | Commercial fishing, aquaculture and other products, tourism |
| Production function (dose-response) | Estimate value as an input in production. Trace impact of physical change of an ES on human welfare. | Data on change in service and impact is often unavailable, only applicable to use values. | Producer and consumer surplus | Sea defence, erosion control, healthy climate, fishing and other products |

| | | | | |
|---|---|---|---|---|
| Replacement cost | Costs of replacing the function with an alternative (manmade) technology or restoration of the ecosystem | Only applicable to use values, tends to overestimate value | | Sea defence, erosion control, healthy climate |
| Defensive/preventive expenditure method, avoided damage costs | Costs and expenditures incurred in avoiding damages of reduced environmental functionality | Only applicable to use values, substitutability issues, typically lower bound estimate, problematic when goods are produced jointly | | Sea defence, erosion control, landscape and biodiversity, health, etc. |
| Market prices | Accounting procedure applicable to market traded goods. | Assumes perfect markets, only possible for private goods, lower bound estimates. | Total market revenue of goods or services | Commercial fishing, aquaculture, harvesting of products |

is close to an ecological threshold, where a small change in one aspect of the system may lead to sharp changes in ecosystem service provision. Therefore, average values, such as average values per ha (total benefit flow divided by total area), should be used with caution (see Brander et al. 2012 for a discussion). Values expressed in different units will also require a different aggregation process; some values may be aggregated over the relevant area, whereas others are to be aggregated over the relevant population. When marginal values are not proportional to unit area, aggregation errors may arise.

In policy appraisal, the results of new or existing valuation studies can be used. The use of existing primary studies, where the valuation results of a 'study site' are used the estimate welfare changes at the 'policy site', is called benefit transfer. Value transfers across different areas can result in errors because of differences in socio-demographic and economic characteristics of the population, as well as ecological and biophysical characteristics of the study sites (Brouwer 2000). These errors can be partly reduced by adjusting for income differences (Bateman et al. 2011). Transfer studies should therefore ideally rely on multiple studies, either through meta-analytical function transfers (e.g. Brander et al. 2012) or by providing a range transfer estimates using different primary studies (e.g. Luisetti et al. 2011 for carbon). Transfers within the same continent and climatic zone (e.g. studies from North- and West-Europe) could be applied with the necessary caution, whilst studies from elsewhere should probably not be applied due to large differences in cultural, economic and ecological differences.

Transfer errors may also arise when studies are transferred over time. One of the fundamental assumptions in BT studies is that preferences underlying WTP estimates are robust over time (Brouwer 2006). However, changes in respondents' socio-economic characteristics or other contextual factors, may alter preferences. When transferring value estimates or functions, underlying preferences are assumed to remain stable. In practice, study results have been transferred over long time periods to estimate the benefit values of ecosystem services at new policy sites. Empirical tests of temporal stability of SP studies for environmental goods and benefits based on CV studies indicate that choices are roughly consistent within short time periods (e.g. 1 year), but may change over longer periods of time. The same results have been found in health care studies when testing the transferability of CE results. We are not aware of any test-retest studies to test temporal stability of other valuation methods. However, for practical reasons related to policy information provision, an arbitrary cut off will necessary in benefit transfer exercises to be able to provide at least some estimates of ecosystem service benefits.

The suitability of benefit transfer also depends on the policy-issue at stake (i.e. how the value estimates will be used): the scale (local, regional, national), the required level of accuracy, the dominant ecosystem services and associated value types (use vs non-use values, cultural values) as well as the available budget for primary data collection. For example, for national assessments such as the UK NEA broadly generalisable values are appropriate, whilst for more regional policy issues such as managed realignment where local sensitivity is important to consider, a more targeted assessment among local key stakeholders may be more useful. Where cultural assets and spiritual values are relevant, social impact analysis through wellbeing and 'shared value' assessments may require other methods, such as deliberative approaches and citizen/stakeholder forums.

Whatever methodology is used to conduct the assessment, all results should be subjected to a rigorous uncertainty and sensitivity analysis. Uncertainty is present at all stages of the assessment process (see Chap. 8), whether it be uncertainty about the magnitude of physical impacts and their geographical and temporal distribution, or uncertainty over the value of changes in ecosystem benefits and goods. Sensitivity analysis allows this uncertainty to be explored in a constructive manner and can be used to identify the parameters of the system which are particularly subject to uncertainty and that have a significant impact on the overall outcome of the assessment.

## 4.6 Shared Values

It is important to note that the value of nature concept is usually interpreted in economic analysis in terms of individuals and their preferences and motivations. The value concept can also however be viewed in a collectivist way, and expressed or elicited in a collective way (i.e. shared values, see UK NEAFO 2014). Cultural or societal values, as well as communal and group values, include principles and values as well as a shared sense of what is worthwhile held by members of a society, community or group. This is in terms of motivations and preferences assigned to groups and culturally transmitted and assimilated over time as social norms. These shared values may not be capable of meaningful monetary expression, but nevertheless they significantly signal that human well-being and quality of life is a function of both individual wants satisfaction and the meeting of a variety of social, health related and cultural collective needs. Cultural values therefore include shared values fostered by and within 'groups' influenced by social rights and wrongs and often acquired over long periods of time and often connected to specific local places.

They may differ in intention from purely self-regarding interest to include other-regarding concerns and therefore encompass a consideration of the ethical arrangements which guide society's concern for nature, place, landscape and seascape, and include motivations such as altruism, bequest value and existence value (Fish et al. 2011). Some arts and humanities analysts would also see aesthetic considerations as an additional value dimension. Society's acceptance of the reliability and legitimacy of decision making processes that have been informed by technical DSSs and have highlighted trade off dilemmas can in certain contexts be heavily influenced by whether shared values have or have not been explicitly recognised and accounted for in the political process.

Shared values often have to be elicited through an interpretative approach which relies on qualitative expressions of value e.g. through the interpretation of documents and media, but also via group discussion, learning and deliberation. Key techniques are deliberative (non-)monetary valuation and participatory multi-criteria analysis (MCA), which hold much promise in terms of a systematic and integrated treatment of utilitarian and other ethical positions, as well as aesthetic considerations. Systematic large scale surveys (e.g. Potts et al. 2011) can begin to unwind broad social values and inform further analysis. Deliberative methods remain, however, at an experimental stage of evolution in environmental management. It is important to note that while techniques are evolving to better understand shared values, the social learning mechanisms are 'processes to be engaged in' facilitating policy deliberation among equal partners (Potts et al. 2011).

## 4.7 Discounting, Ethics, Equity and Distributional Considerations

It is often necessary to choose between options that differ in temporal patterns of costs and benefits, or that differ in their duration. Discounting provides a common matrix that enables comparison of costs and benefits that occur at different points in time. Use of discounting yields an outcome in which future costs and benefits are valued less highly than those that occur in the present, and the procedure is integral to CBA and Cost Effectiveness Analysis (CEA). The choice of the discount rate can have a significant effect on the economic viability of management options and their relative economic ranking. It signals the rate at which future consumption is to be traded against consumption in the present. Use of a 'high' positive rate of discount discriminates against the future and, in project terms, against options that involve high initial costs and a stream of benefits that extends far out into the future (e.g. coastal wetland creation, restoration, or maintenance within a coastal defence or protection strategy). Instead it tends to favour projects that have immediate benefits and delayed cost burdens (Turner 2007). But while a low discount rate favours the future, this may be politically and morally questionable if immediate wellbeing increases are slowed or compromised altogether and the burden falls disproportionately on the poor.

The discounting question raises a number of much deeper ethical and strategic considerations related to equity and fairness principles and practice. Fairness in contemporary society (intra-generational equity) is sidestepped in conventional applications of CBA via the acceptance of the economic efficiency criterion which weights all benefits and costs equally, regardless of whether they affect rich or poor in society (known as Potential Pareto Improvement as determined by the Hicks-Kaldor compensation test) (Gowdy 2004; Turner 2007). We have made a case for actual compensation (financial and environmental), especially given the 'contested' nature of environmental change in coastal zones, in Chap. 2 and the Balance Sheets Approach. The debate around discounting has a long history and involves some difficult ethical questions; we summarise some of this in next.

We first focus on fairness across time (intergenerational equity) and the practice of discounting. The standard CBA practice of positive, fixed and short term (<25 years) discounting does not sit easily within policy contexts with pressures and drivers such as climate change and related global economic forces. A growing number of analysts agree that discounting at a constant and relatively high (i.e. determined by reference to market interest rates data) rate of discount over time horizons of 100 years or more is problematic. The effect is to make even large costs or benefits incurred in the distant future seem inconsequential and this feels intuitively wrong (Weitzman 1998). There seems to be a tyranny imposed by current generations on the future when, for example, there is current inaction conditioned by cost considerations and a neglect of low weighted future benefits, e.g. climate change (Groom et al. 2005). Asheim (2012) has summarised the dilemma in the following way: contemporary society needs to distinguish between what the current generations as a collective should do ethically to serve the interests of all generations from an impartial perspective, and what contemporary countries or individuals should do strategically to serve their own interests when such actions influence the strategic action of other countries and individuals.

The ethical and strategic dilemma is not as straightforward as it may appear upon superficial examination. Zero or negative discounting poses a threat to the least well off in today's society and can result in large sacrifices from the present for the benefit of later generations who may be better off, for example, because of innovation. A single invariant low rate of discount could in some circumstances allow a greater volume of projects to pass the CBA test and therefore deplete non-renewable stocks or cause pollution. Nevertheless, some modifications to the standard CBA discounting procedure have been adopted, for example in UK public sector project appraisal (HMT 2003). A time declining discount rate (DDR) procedure over at least 100 years is now recommended for projects with significant environmental impacts.

A range of reasons have been put forward in support of DDR, including uncertainty about future interest rates and the macro-economic state of the economy (Weitzman 1998; Gollier 2002). Some empirical evidence exists for 'hyperbolic discounting' indicating that individuals value medium and distant futures on an equivalent basis, i.e. the discount rate falls the longer the time horizon (Henderson and Bateman 1995). It may be that individuals live in relative not absolute time and therefore revise and re-evaluate plans continually as time passes; or over time individuals pass through different stages of life and change as people (Henderson and Bateman 1995; Heal 1998; Frederick et al. 2002). Advocates of the conventional discounted utilitarian approach in conventional CBA would, however, counter that social discounting as practiced by governments should not mimic the 'time inconsistent' or 'irrational' behaviour of individuals exhibiting hyperbolic discounting behaviour. But while policy inconsistency at a given period of time is an institutional failure and should be corrected, policy switching over longer periods of time are inevitable and necessary if future uncertainties and 'surprises' are unavoidable. Finally, Knetsch (2005) has claimed that individuals discount future losses at a lower rate than the value of future gains and that therefore rates reflecting observed individual preferences would give more weight to future environmental losses, justify greater current sacrifices to deal with them and support policies that reduce the risk of future loss.

The ethics versus strategic behaviour dichotomy is the focus for a key set of arguments. If individual people also have (individual/shared) other-regarding (social) preferences and if they trade off their own material interests against the wider interests of society, then some notion of fairness that captures social preferences is required. Roemer (2011) has argued that the utilitarian social welfare function used in conventional CBA assumes that the decision problem for a society with many generations is equivalent to the decision problem of an infinitely-lived consumer. This claim is refuted by Dasgupta (2011), who prefers conventional social welfare functions, but both agree that the discount rate based on market data and applied to the climate change problem is too large. They disagree on what is the better ethically defensible sustainability criterion, with Roemer favouring the Rawlsian intergenerational maximin approach, i.e. each generation passes on a non-declining, in per capita welfare terms, capital (human, physical and natural capital components) bequest. Roemer argues that consumption as conventionally defined in economics is not the only component of welfare or wellbeing. Educated leisure, quality of the local to global environment and knowledge are also direct inputs into welfare. Intergenerational maximin is not problem free (e.g. how much sacrifice is a fair burden for the current generation rich and poor?), but in the spirit of moral pluralism (i.e. there is no meta-ethical criterion and the context and consequences of ethical choices should not be ignored) intergenerational maximin may still prove to be a usable ethical guide.

Finally, Asheim (2012) has proposed what he calls an equity-rank-discounted utilitarian intergenerational equity position in which welfare is discounted not according to time but according to rank. This approach it is claimed can combine equal treatment of generations with social discounting by giving priority for the worse off not only due to their absolute level of wellbeing (Rawlsian-maximin) but also their relative rank in wellbeing. If the future is better off than the present, then this criterion is on a par with discounted utilitarianism. However, if, for example, climate change brings an end to the past positive correlation between time and increasing welfare, then rank-discounted utilitarianism makes a greater call for present action (and lower discount rates) to protect the interests of future generations. In the marine environment, scientific work has shown that key processes are slow with timescales over 1,000 years or more, for example, with ice sheet or deep ocean changes. So changes in current policy related to economic development and climate change could have impacts stretching out 1,000 years (Stouffer 2012). But will this make the future worse off than contemporary society? Taking a precautionary approach, the 2007 IPCC (Intergovernmental Panel on Climate Change) sea level rise predictions (maximum 2 ft rise in sea level by the end of the century) now seem too optimistic as they failed to factor in ice sheet melting impacts. Now some estimates put the sea level rise up to 7 ft (Young and Pilkey 2010), and some coastal authorities have design plans with 2.5 ft (in the Netherlands) and 4.6 ft (in California, USA) rise parameters built into them to correct for the low IPCC 2007 estimates (*ibid.*). While ice sheet melting is non-linear and difficult to predict, the threat posed to human welfare is significant. Low elevation coastal zones i.e. contiguous areas with elevations below 10 m, contain 10 % of the global population and have expanding populations, and a large proportion of the world's megacities.

## 4.8 Multi-criteria DSSs

Most methods of economic assessment are concerned with determining the efficiency of policy options, where efficiency is defined in an economic sense in which the most efficient solution is the one that increases overall welfare to the greatest extent. But as we argued earlier efficiency is not necessarily associated with equity (i.e. questions of where welfare benefits or costs fall; e.g. on particular sectors of industry, certain social classes, certain geographical areas, etc.). However, sustainable solutions must consider both equity and efficiency. Given the 'contested' nature of coastal socio-ecological resource systems (Ostrom 2007), questions of trade-offs, social justice, equity and compensation are likely to loom large in public debate.

Appropriate DSSs can therefore be informed, for example, by a better understanding of relevant social and policy networks (Bodin and Crona 2009; Borgatti et al. 2009; Bainbridge et al. 2011); and also via methods and techniques encompassing multiple values and decision criteria. Economic assessment methodologies can be modified to incorporate equity issues (e.g. via the application of weights to costs and benefits), and the economic analysis itself can be augmented by a wider trade-off analysis, for example using MCA or deliberative (non-)monetary valuation techniques.

DSSs and their component methods and tools such as CBA, CEA and MCA, require the acceptance of different assumptions about the capacities and motivations of the individuals involved, and the role the methods play in framing/scoping the assessment process. From an institutional perspective, CBA and other methods can be characterised as value articulating institutions, in the sense of rule structures facilitating value (Vatn 2009). If the existence of plural rationalities is accepted, the role of such institutions is to signal which rationality (economic or otherwise) is expected. The choice of assessment approach, for example, is related to whether the benefits and costs involved are linked to an exchangeable commodity or some kind of public good; if it is the allocation of a public type good that is being contested then deliberative methods may play a useful role, although they need to be tailored to suit the situation, e.g. citizens or stakeholder forum or a hybrid. The Balance Sheets Approach in Chap. 2 aims to provide an appraisal format based on a pluralistic perspective.

The particular sequencing of policy tools (methods and tools) through the Balance Sheets Approach starts with CBA or CEA but then encompasses other complementary 'tools' to apply AM principles in coastal management. In the next chapter we take a closer look at indicators for coastal ecosystem services.

## References

Asheim, G. B. (2012). Discounting while treating generations equality. In R. W. Hahn & A. Ulph (Eds.), *Climate change and common sense: Essays in honour of Tom Schelling*. Oxford: Oxford University Press.

Bainbridge, J. M., Potts, T., & O'Higgins, T. G. (2011). Rapid policy network mapping: A new method for understanding governance structures for implementation of marine environmental policy. *PLoS ONE, 6*, e20149.

Bateman, I. J., Ferrini, S., Brouwer, R., Schaafsma, M., Barton, D. N., Dubgaard, A., Hasler, B., Hime, S., Liekens, I., Navrud, S., De Nocker, L., Sceponavicitue, R., & Semeniene, R. (2011). Making benefit transfers work: Deriving and testing principles for value transfers for similar and dissimilar sites using a case study of the non-market benefits of water quality improvements across Europe. *Environmental and Resource Economics, 50*, 368–387.

Baveye, P. C., Baveye, J., & Gowdy, J. (2013). Monetary valuation of ecosystem services: It matters to get the timeline right. *Ecological Economics, 95*, 231–235.

Bodin, Ö., & Crona, B. (2009). The role of social networks in natural resource governance: What relational patterns make a difference? *Global Environmental Change, 19*, 366–374.

Borgatti, S. P., Mehra, A., Brass, D. J., & Labianca, G. (2009). Network analysis in the social sciences. *Science, 323*, 892–895.

Brander, L. M., Florax, R. J. G. M., & Vermaat, J. E. (2006). The empirics of wetland valuation: A comprehensive summary and a meta-analysis of the literature. *Environmental and Resource Economics, 33*, 223–250.

Brander, L. M., Brauer, I., Gerdes, H., Ghermandi, A., Kuik, O., Markandya, A., Navrud, S., Nunes, P. A. L. D., Schaafsma, M., Vos, H., & Wagtendonk, A. (2012). Using meta-analysis and GIS for value transfer and scaling up: Valuing climate change induced losses of European wetlands. *Environmental and Resource Economics, 52*, 395–413.

Brouwer, R. (2000). Environmental value transfer: State of the art and future prospects. *Ecological Economics, 32*, 137–152.

Brouwer, R. (2006). Do stated preference methods stand the test of time? A test of the stability of contingent values and models for health risks when facing an extreme event. *Ecological Economics, 60*(2), 399–406.

Costanza, R., D'Arge, R., Groot, R. D., Farber, S., Grasso, M., Hannon, B., Limburg, K., Naeem, S., O'Neil, R. V., Paruelo, J., Raskin, R. G., Sutton, P., & Belt, M. V. D. (1997). The value of the world's ecosystem services and natural capital. *Nature, 387*, 253–260.

Costanza, R., de Groot, R., Sutton, P., van der Ploeg, S., Anderson, S. J., Kubiszewski, I., Farber, S., & Turner, R. K. (2014). Changes in the global value of ecosystem services. *Global Environmental Change, 26*, 152–158.

Dasgupta, P. (2011). The ethics of intertemporal distribution in a warming planet: Reply and response to J.E. Roemer. *Environmental and Resource Economics, 50*, 475–493.

Fish, R., Burgess, J., Chilvers, J., Footitt, A., Haines-Young, R., Russel, D., Turner, K., & Winter, D. M. (2011). *Participatory and deliberative techniques for embedding an ecosystems approach into decision making*. Full Technical Report, Project Code: NR0124.

Frederick, S., Loewenstein, G., & O'Donoghue, T. (2002). Time discounting and time preference: A critical review. *Journal of Economic Literature, 40*, 351–401.

Gollier, C. (2002). Discounting an uncertain future. *Journal of Public Economics, 85*, 149–166.

Gowdy, J. M. (2004). The revolution in welfare economics and its implications for environmental valuation and policy. *Land Economics, 80*, 239–257.

Groom, B., Hepburn, C., Koundouri, P., & Pearce, D. W. (2005). Declining discount rates: The long and short of it. *Environmental and Resource Economics, 32*, 445–493.

Heal, G. (1998). *Valuing the future: Economic theory and sustainability*. New York: Columbia University Press.

Heal, G. (2000). *Nature and the marketplace*. Washington, DC: Island Press.

Henderson, N., & Bateman, I. (1995). Empirical and public choice evidence for hyperbolic social discount rates and the implications for intergenerational discounting. *Environmental and Resource Economics, 32*, 13–34.

HMT. (2003). *Green book: Appraisal and evaluation in central government*. London: HMSO.

Howarth, R. B., & Farber, S. (2002). Accounting for the value of ecosystem services. *Ecological Economics, 41*, 421–429.

Knetsch, K. (2005). Gains, losses and the US-EPA economic analyses guidelines: A hazardous product? *Environmental and Resource Economics, 32*, 91–112.

Luisetti, T., Turner, R. K., Bateman, I. J., Morse-Jones, S., Adams, C., & Fonseca, L. (2011). Coastal and marine ecosystem services valuation for policy and management: Managed realignment case studies in England. *Ocean and Coastal Management, 54*, 212–224.

Luisetti, T., Jackson, E., & Turner, R. K. (2013). Valuing the European 'Coastal Blue Carbon' storage benefit. *Marine Pollution Bulletin, 71*(1–2), 101–106.

Millennium Ecosystem Assessment (MEA). (2005). *Millennium ecosystem assessment: Ecosystems and human wellbeing: Synthesis*. Washington, DC: Island Press.

Morse-Jones, S., Bateman, I. J., Kontoleon, A., Ferrini, S., Burgess, N. D., & Turner, R. K. (2012). Stated preferences for tropical wildlife conservation amongst distant beneficiaries: Charisma, endemism, scope and substitution effects. *Ecological Economics, 78*, 9–18.

Norgaard, R. B. (2010). Ecosystem services: From eye-opening metaphor to complexity blinder. *Ecological Economics, 69*(6), 1219–1227.

Ostrom, E. (2007). A diagnostic approach for going beyond panaceas. *Proceedings of the National Academy of Sciences, 104*, 15181–15187.

Pearce, D. W., & Turner, R. K. (1990). *Economics of natural resources and the environment*. Hemel Hempstead: Harvester Wheatsheaf.

Potts, T., O'Higgins, T., Mee, L., & Pita, C. (2011). *Public perceptions of Europe's seas – A policy brief*. Eu Fp7 Knowseas Project. ISBN 0-9529089-3-X.

Roemer, J. E. (2011). The ethics of intertemporal distribution in a warming planet. *Environmental and Resource Economics, 48*, 363–390.

Stouffer, R. J. (2012). Oceanography: Future impact of today's choices. *Nature Climate Change, 2*, 397–398.

Turner, R. K. (1999). The place of economic values in environmental valuation. In I. J. Bateman & K. G. Willis (Eds.), *Valuing environmental preferences* (pp. 17–41). Oxford: Oxford University Press.

Turner, R. K. (2007). Limits to CBA in UK and European environmental policy: Retrospects and future prospects. *Environmental and Resource Economics, 37*, 253–269.

Turner, R. K., Paavola, J., Cooper, P., Jessamy, V., & Georgiou, S. (2003). Valuing nature: Lessons learned and future research directions. *Ecological Economics, 46*, 493–510.

Turner, R. K., Morse-Jones, S., & Fisher, B. (2010). Ecosystem valuation: A sequential decision-support system and quality assessment issues. *Annals of the New York Academy of Sciences, 1185*, 79–101.

UK National Ecosystem Assessment Follow-on. (2014). *The UK national ecosystem assessment follow-on: Synthesis of the Key findings*. Cambridge, UK: UNEP-WCMC/LWEC.

UNU-IHDP and UNEP. (2012). *Inclusive wealth report 2012. Measuring progress toward sustainability*. Cambridge, UK: Cambridge University Press.

Vatn, A. (2009). An institutional analysis of methods for environmental appraisal. *Ecological Economics, 68*, 2207–2215.

Weitzman, M. (1998). Why far distant future should be discounted at its lowest possible rate. *Journal of Environmental Economics and Management, 26*, 200–209.

Young, R., & Pilkey, O. (2010). How high will seas rise? Get ready for seven feet. *Yale Environment 360*. 14 Jan 2010.

# Part II
# Practice

# Chapter 5
# Identification of a Practicable Set of Ecosystem Indicators for Coastal and Marine Ecosystem Services

**Jonathan P. Atkins, Daryl Burdon, and Michael Elliott**

## 5.1 Introduction

Given the complexity of the coastal and marine ecosystem and the need for integrated management, indicators are required to provide insight into the behaviour and state of coastal and marine ecosystems, together with an indication of the trajectory of change due to natural and human events (Elliott 2011; Gibbs 2012; TIDE 2013). In this chapter we examine what is meant by indicators and explore their key purposes and application. Following on from the UK NEAFO ecosystem services framework described in Chap. 1, we identify a set of ecosystem service indicators for components and processes, intermediate services, final services and goods/benefits applicable in a practicable way to UK coastal and marine systems. These ecosystem service indicators reflect the *State changes* and *Impacts* in the DPSIR (*Drivers-Pressures-State changes-Impacts-Responses*) framework (Atkins et al. 2011) (see Chap. 2). Examples of national-level data sources available to support indicator use for the UK coastal and marine environments are identified. A more detailed consideration is provided of the application of the indicators to fisheries and aquaculture, and to carbon sequestration and storage. Case studies are also discussed which demonstrate the importance of site-specific data sources in relation to marine protected areas and to managed realignment sites. An operational assessment of indicators specifically involving the specification of targets, for example for compliance purposes, is beyond the scope of this chapter.

J.P. Atkins (✉)
Hull University Business School, University of Hull, Hull HU6 7RX, UK
e-mail: J.P.Atkins@hull.ac.uk

D. Burdon • M. Elliott
Institute of Estuarine and Coastal Studies, University of Hull, Hull HU6 7RX, UK



R.K. Turner, M. Schaafsma (eds.), *Coastal Zones Ecosystem Services*,
Studies in Ecological Economics 9, DOI 10.1007/978-3-319-17214-9_5

## 5.2 Understanding Indicators and Their Key Purposes

An indicator can be described as a measure or metric based on evidence that can be verified and that conveys information about more than just itself (UNEP-WCMC 2009). Gabrielsen and Bosch (2003) suggested an indicator '*is an observed value representative of a phenomenon of study. In general, indicators quantify information by aggregating different and multiple data. The resulting information is therefore synthesised. In short, indicators simplify information that can help to reveal complex phenomena*'. More specifically, indicators can be of two types: firstly, ecosystem indicators are '*measures of key ecosystem properties reflecting changes in ecosystem services and can provide information on the direction and possible magnitude of the impact or response of an ecosystem to stress*' (van den Belt and Costanza 2011); secondly, an indicator can be a quantitative value against which change is measured and where the value to be exceeded is incorporated in a statutory or policy instrument, where compliance with it is judged by monitoring (McLusky and Elliott 2004; Elliott 2011). While indicators can reflect state, trends and/or performance of the marine system, they can also reflect the marine ecosystem natural capital stocks and the flow of ecosystem services of significant value (benefits) to human society. Hence, indicators can reflect the state of the science of an area or provide management tools.

Aubry and Elliott (2006) suggested environmental indicators should have three basic functions:

- to simplify: amongst the diverse components of an ecosystem, a few indicators are needed according to their perceived relevance for characterising the overall state of the ecosystem;
- to quantify: the indicator is compared with reference values considered to be characteristic of either 'pristine' or heavily impacted ecosystems to determine changes from reference or expected conditions; and
- to communicate: with stakeholders and policy makers, by promoting information exchange and comparison of spatial and temporal patterns.

The first of these recognises that a key challenge in the development and use of ecological indicators is to determine '*which of the numerous measures of ecological systems characterise the entire system yet are simple enough to be effectively and efficiently monitored and modelled*' (Dale and Beyeler 2001). Indicator choice needs then to be grounded within a conceptual framework such that the individual indicator characteristics are not overemphasised as formal selection criteria and greater attention is given to the function of the indicator within an analytical problem-solving logic. Indicators selected relating to cause and effect should have a function recognising the interrelations and causality within the environmental system (Niemeijer and de Groot 2008). Although not translated into ecosystem services, Aubry and Elliott (2006) emphasised that to cover the pressures (as activities), the hydromorphological change due to those pressures, and the effects (State change and Impacts) of those pressures a weighted suite of indicators should be used.

Kandziora et al. (2012), recognising the complexity of the human-environmental system, presented a suite of ecological indicators, based on the Millennium Ecosystem Assessment's ecosystem services (MEA 2005), that capture the interrelations between ecosystem properties, biodiversity, ecosystem integrity, ecosystem services and human welfare. Their study concluded that ecosystem service indicators meet the criteria of being adequate human-environmental system indicators and, therefore, are an appropriate instrument for decision making and management.

On the second function above, it is axiomatic that management of the marine system requires measurement and that monitoring is required to provide those measurements (Elliott 2011). In the detection of change, those monitoring measures have to be against a desired outcome, for example a baseline, reference condition, trigger or threshold value (Gray and Elliott 2009), and ideally an action is defined a priori before the indicators and monitoring are employed. For example, an industry would be given a trigger value level in a licence and then action would be taken if that level is breached. Each of these monitoring measures are then indicators which highlight a deviation from change; for example, the EU Water Framework Directive (WFD), Marine Strategy Framework Directive (MSFD), Habitats and Species Directive and Environmental Impact Directive are all based on a knowledge of what an area should be like (its 'normal' or reference condition) and whether it has deviated (or will in the future deviate) from this due to human activities. Hence there is a need for indicators to determine the state of that normal condition, the degree of deviation and the trajectory of that change. This is inherent in the European Environment Agency's typology for indicators which classifies indicators into four types according to which of the following questions they address: 'What is happening to the environment and to humans?' (Type A or descriptive indicators); 'Does it matter?' (Type B or performance indicators); 'Are we improving?' (Type C or efficiency indicators), and 'Are we on the whole better off?' (Type D or total welfare indicators). Smeets and Weterings (1999) in their presentation of the EEA typology recognised that these indicators necessarily include 'red-flag', tipping point or threshold indicators and early warning indicators as an aid to management.

On the third function, communicating the compliance with or deviation from, for example, a baseline is a further challenge for indicator use. In the present context, ecosystem service indicators are by their nature inherently interdisciplinary and so finding a language common to all stakeholders is not straightforward, particularly when combining different philosophies, paradigms and research techniques. A common language is also required to communicate objectives, methods and outcomes to a number of different audiences, from the lay-person through to specialists and policy-makers (UNEP-WCMC 2011).

The literature refers to the SMART characteristics of indicators, which follows from the work of Doran (1981). According to this set of criteria, in order to be operational, valuable and successful, the management of the environment requires indicators which are Specific, Measurable, Achievable/Appropriate/Attainable, Realistic/Results focused/Relevant, and Time-bounded/Timely. Without meeting these five criteria, it is suggested that the indicators cannot be used in measuring, monitoring and managing change (e.g. Dauvin et al. 2008; Gray and Elliott 2009).

As an extension of this, Elliott (2011) defines 18 required properties of indicators and monitoring parameters specifically for successful marine management, a number of which were additional to the SMART characteristics. These additional properties included: anticipatory; biologically/environmentally important; broadly applicable and integrative over space and time; giving continuity over time and space; cost-effective in monitoring; grounded in theory/relevant and appropriate; interpretable; low redundancy; non-destructive; responsive feedback to management; sensitive to a known stressor or stressors; and socially relevant. In linking these to the Pressures, State change and Impact elements of the DPSIR as a management framework, Elliott (2011) argues these 18 attributes need to be fulfilled in the selection of coastal and marine indicators.

Turning to the literature that has attempted to identify indicators for ecosystem services, of which there are several, few have specifically identified indicators for changes of direct relevance to the UK coastal and marine environment. On those that are more general, UNEP-WCMC (2011) assessed ecosystem indicators for four categories of ecosystem services (provisioning, regulating, supporting and cultural) based on evidence from 34 sub-global assessment reports. The study considered a wide range of ecosystem types including coastal, cultivated, dryland, forest, inland water, island, marine, mountain and urban regions. Indicators were identified for 23 ecosystem services following the generic TEEB (The Economics of Ecosystems and Biodiversity) framework (Kumar 2010). A distinction was made between state indicators (how much of the service is present) and performance indicators (how much can be used/provided in a sustainable way) following de Groot et al. (2010a, b). The study concluded that indicators of ecosystem services were underdeveloped and failed to convey a complete picture, with the average quality of ecosystem service indicators and data availability being considered poor, particularly if the indicators were to be used in combination as a decision support tool (indicators of cultural, supporting and regulating services being especially problematic).

Liquete et al. (2013) used a meta-analysis to systematically review the current status and future prospects for the assessment of marine and coastal ecosystem services. They identified 145 papers which specifically assess marine and coastal ecosystem services, mainly with a focus on mangroves and coastal wetlands in Europe and North America. A catalogue of 476 ecosystem service indicators was created and gaps identified in current knowledge. Most indicators relate to a limited set of ecosystem services and benefits, including food provision (fisheries), water purification, coastal protection and recreation/tourism. This systematic review established a background that can facilitate the planning and integration of future ecosystem service assessments in the coastal and marine environment.

Both Böhnke-Henrichs et al. (2013) and Hattam et al. (2015) provided sets of ecosystem service indicators within the TEEB framework specifically for application in the European marine environment. Böhnke-Henrichs et al. argued that the ecosystem service concept has rarely been applied to marine planning and management due to a lack of a well-structured, systematic classification and assessment of marine ecosystem services. Given this, their study proposed such a typology and provided guidance on the selection of indicators in relation to marine spatial planning and management. Hattam et al. linked ecosystem service indicators to their adaptation of the TEEB framework by identifying the need to distinguish between indicators of ecosystem services that are entirely ecological in nature, indicators for the ecological processes contributing to the delivery of these services, and indicators of benefits that reveal the realised human use or enjoyment of an ecosystem service. For the purpose of identifying indicators, the TEEB framework is structurally different to that outlined in Chap. 2.

## 5.3 Ecosystem Indicators for Coastal and Marine Ecosystem Services

The ecosystem services framework, developed for the UK National Ecosystem Assessment (UK NEA 2011), recognises the importance of distinguishing between basic processes, intermediate services and final services, and goods/benefits (see Chap. 2). Given that the framework was designed to be generally applicable across ecosystems, in order to increase its relevance to the coastal and marine environment the framework was modified under the NERC-funded Valuing Nature Network Coastal Management project (Turner et al. 2013) and again following UK NEAFO WP3b (Marine Economics) workshops (see Fig. 2.6). To capture the diversity and complexity of the coastal and marine systems the indicators need not only to be specific to ecosystem services but also to relate to the components and processes and goods/benefits as identified within the ecosystem services framework. Using indicators of coastal and marine components and processes to detect changes in ecosystem service provision may be viewed, ecologically, as a bottom-up approach, while application of indicators for the resulting goods/benefits to detect change in ecosystem service provision may be regarded as a top-down approach.

Based on the UK NEAFO WP3b framework, Table 5.1 provides a range of ecosystem service indicators identified for each category, and these have been subject to a process of review for their ability to provide insight into the behaviour, state and trajectory of the coastal and marine systems by the WP3b project group at two workshops. Given the requirement for ecosystem service indicators to provide such insights, some of the examples in the table are expressed as levels (quantity, quality, etc.) to reflect the state of the ecosystem at a given point in time whereas others are expressed as changes to reflect the trajectory and/or behaviour of ecosystem service provision over time. The final selection of the most practicable indicator(s) will depend on the context and operational needs. All of the indicators identified are expressed in natural science units or units having more anthropocentric relevance; indicators measured in monetary units are discussed in Chap. 6. The indicators listed within the table are examples and not meant to be exhaustive of all such indicators. Moreover, it is likely that such indicators will need to be formulated in a more specific way for application in particular cases or for particular management purposes. The indicators reflect state, trends and/or performance within the coastal

**Table 5.1** Indicators of coastal and marine ecosystem components and processes, intermediate services, final services, and goods/benefits. Examples of UK data sources are provided

| Coastal and marine ecosystems | Example indicators (units) | Examples of UK data sources |
|---|---|---|
| **Components and processes** | | |
| Habitats and species | Abundance (number); biomass (g, kg); species diversity (Shannon Wiener Index); % cover of habitat; area of habitat (ha); gene pool; biotope matrix; AMBI (marine biotic index); phytoplankton index | MESH Atlantic (2004–2008, 2010–2014); UK SeaMap (2006, 2010); European Marine Ecosystem Observatory; Defra MB0102 data layers; OBIS SEAMAP |
| Sea space | Area of surface (ha); volume ($m^3$); tidal range (m); depth (m); bathymetry; topography | UK SeaMap (2006, 2010); Scotland's Marine Atlas; Scotland's Marine Plan Interactive; SeaZone Solutions |
| Sea water | Depth (m); volume ($m^3$); pH; salinity; turbidity (mg/l) | UK SeaMap (2006, 2010); Defra DEM bathymetry maps; Scotland's Marine Atlas; Scotland's Marine Plan Interactive; SeaZone Solutions |
| Substratum | Area (ha) and depth (m) by type (mud, sand, gravel, etc.) | UK SeaMap (2006, 2010); MESH Atlantic (2004–2008, 2010–2014) |
| Production | Community production (kcal); net productivity by species (kcal/ha/yr); P:B (productivity:biomass) ratios | EU MyOcean project; published and grey literature |
| Decomposition | Amount and number of decomposers (n/ha); decomposition rate (kg/ha/yr) | Site-specific published and grey literature |
| Food web dynamics | Changes over time in community composition (abundance (number); biomass (g, kg); species diversity (diversity indices)); population dynamics (age classes, male:female ratios) | DASSH website; fish trawl surveys database (ICES, 1989 to present); site-specific published and grey literature |
| Ecological interactions | Competition for food and space; resilience and resistance (predator:prey, adults:juveniles, etc.) | Site-specific published and grey literature |
| Hydrological processes | Current speed (m/s) and direction; wave height; changes in temperature (°C); changes in salinity; changes in turbidity (mg/l); NAO (North-Atlantic Oscillation) cycles | Defra MB0102 data layers; UK SeaMap (2006, 2010); EU Global Ocean OSTIA Sea surface temperature and sea ice analysis REPROCESSED (1985–2007) |
| Geological processes | Sediment accumulation rates; beach slopes and gradients; seabed form; channel depths; erosion-deposition cycles | UK SeaMap (2006, 2010) |

| | | |
|---|---|---|
| Evolutionary process | Changes in genetic diversity; mutation rates; influx/efflux of species (number) | Site-specific published and grey literature |
| **Intermediate ecosystem services** | | |
| Primary production | Quantity of primary production (g C per unit area/volume); quality of primary production (e.g. efficiency of converting sunlight to carbon) | Site-specific published and grey literature |
| Larval and gamete supply | Quantity of larvae/gametes supplied to a particular location (number per $m^3$); quality of larvae/gametes supplied to a particular location (% affected by disease; mortality rates) | UK spawning and nursery grounds (Ellis et al. 2012); fish eggs and larvae database (ICES, 1967-present) |
| Nutrient cycling | Changes (output of the system less input to the system) in the amount of nitrates, phosphates, silica (g per unit area/volume); denitrification (kg N/ha/yr) | Site-specific published and grey literature; Cefas modelling data |
| Water cycling | Changes (output of the system less input to the system) in the amount of water ($m^3$) | Relevant Environment Agencies (EA, SEPA, NIEA); site-specific published and grey literature |
| Formation of species habitat | Change in area of habitat (per ha); change in quality of habitat; change in number of juveniles | MESH Atlantic (2004–2008, 2010–2014); UK SeaMap (2006–2010) |
| Formation of physical barriers | Change in amount of natural barriers e.g. saltmarsh, reefs, sand dunes, reed beds etc. (% cover, ha) | MESH Atlantic (2004–2008, 2010–2014); UK SeaMap (2006–2010) |
| Formation of seascape | Changes in area by scenic type (ha, % cover, visual range (m, km)) | MESH Atlantic (2004–2008, 2010–2014); UK SeaMap (2006–2010) |
| Biological control | Quantity of pest/disease/predator-control species (number); quality of pest-control species (prevalence) | DASSH website; published and grey literature |
| Natural hazard regulation | Width or area (and volume if applicable) of saltmarsh, reed bed, mudflat, sand dunes etc. (m, % cover, ha, $m^3$) absorbing energy | MESH Atlantic (2004–2008, 2010–2014); UK SeaMap (2006–2010) |
| Waste breakdown and detoxification | Water quality indicators (N mg/l, P mg/l, bacterial levels mg/l etc.); total dissolved solids (mg/l); water volume; assimilative capacity | Relevant Environment Agencies (EA, SEPA, NIEA) |

(continued)

**Table 5.1** (continued)

| Coastal and marine ecosystems | Example indicators (units) | Examples of UK data sources |
|---|---|---|
| Carbon sequestration | Amount of carbon dioxide sequestered (tonnes of $CO_2$ per $m^2$ or $m^3$); assimilative and recycling capacity, net carbon burial (tonnes per ha per year) | DASSH website |
| **Final ecosystem services** | | |
| Coastal and marine biota | Fish and shellfish population size (biomass of fish/shellfish in tonnes); quality of the fish, shellfish (age profile; length profile; % affected by disease; mortality rates); quantity of seaweed stock (biomass in tonnes, area of seaweed ha); quality of seaweed stock (% affected by disease; mortality rates); quantity of raw material (tonnes); quality of raw material (concentration); quantity of species with potential/actual useful genetic raw material (tonnes); Gene bank composition (e.g. number of species and subspecies); quality of species with potential/actual useful genetic raw material (tonnes equivalent if variation in quality) | UK spawning and nursery grounds (Ellis et al. 2012); national fish population dataset (EA, 2004–2014); fish trawl surveys database (ICES, 1989 to present); DASSH website; site-specific published and grey literature |
| Climate regulation | Greenhouse gas balance especially carbon sequestration (g C); quantity of greenhouse gases fixed and/or emitted; effect on climate parameters (temperature, rainfall, wind, etc.) | Site-specific published and grey literature |
| Natural hazard protection | Width or area of saltmarsh, reed bed, mudflat, sand dunes etc. providing natural hazard protection (m, % cover, ha); sediment stabilisation properties; water retention capacity ($m^3$); (wave) energy dissipation capacity (joules/$m^2$) | UK SeaMap (2006–2010); MESH Atlantic (2004–2008, 2010–2014) |

| | | |
|---|---|---|
| Clean water and sediments | Amount of waste that can be recycled or immobilised (tonnes); Biological oxygen demand (mg $O_2$/litre/day); amount of organic matter in water and sediment (mg/l); amount of heavy metals in water and sediment (mg/l); amount of bacteria in water and sediments (mg/l); heavy metal (and other pollutant) content in marine organisms (concentration) | Relevant Environment Agencies (EA, SEPA, NIEA) |
| Places and seascapes | Number of designated sites; number/area of specific seascape features; % of total natural seascape | JNCC website; English Heritage; National Trust |
| **Goods/benefits** | | |
| Food (wild, farmed) | Nutrition from seafood consumption (g protein/year or g protein/year/head or per household); fish landed for human consumption (landings data at particular times and places in tonnes) | MMO Landings data; FAO Statistics; Office for National Statistics |
| Fish feed (wild, farmed, bait) | Nutrition from non-human seafood consumption (g protein/year); fish landed not for human consumption (landings data at particular times and places in tonnes); bait landed for angling (tonnes); quantity of bait collected by type | MMO Landings data; FAO Statistics; Office for National Statistics |
| Fertilisers and biofuels | Mineral and other content used (e.g. N concentration in g, tonnes); quantity of biomass harvested for energy production | Site-specific published and grey literature |
| Ornaments and aquaria | Ornamental use (tonnes) by type; number of people/businesses who rely on ornamental artefacts (no.) | Site-specific published and grey literature |
| Medicines and blue biotechnology | Contribution to medicines (number of medicines, improvements in mortality rates and quality of life, etc.); total amount of useful substances that can be extracted (kg/ha); quantity of specific blue biotechnologies (e.g. biocatalysts) | Site-specific published and grey literature |

(continued)

**Table 5.1** (continued)

| Coastal and marine ecosystems | Example indicators (units) | Examples of UK data sources |
|---|---|---|
| Healthy climate | Physical damage avoided through net GHG sequestration and effects on climate parameters; bodily harm avoided (lives saved and injuries not incurred) through net GHG sequestration and effects on climate parameters | Site-specific published and grey literature |
| Prevention of coastal erosion | Number of prevented hazards (number per yr); quantity of risk prevention (quantity of assets affected adjusted for risk); amount of man-made infrastructure not required (length/width/height in m) | Local Authorities; National Coastal Erosion Risk Mapping (NCERM) data |
| Sea defence | Amount of man-made infrastructure no longer required; businesses and people protected from flooding; number of flood related mortalities; flooding days per year (combined with rainfall indicator) | Relevant Environment Agencies (EA, SEPA, NIEA); Local Authorities; Environmental Valuation Reference Inventory (EVRI) |
| Waste burial/removal/ neutralisation | Quantity of degradable waste deposited (tonnes by type); quantity of non-degradable waste deposited (tonnes by type); pollution damage avoided by not disposing degradable and non-degradable waste elsewhere (type and extent); treatment and engineering works not required (type and capacity); changes in activity not implemented due to capacity to immobilise waste (quantity and/or other characteristics of activity) | Relevant Environment Agencies (EA, SEPA, NIEA); Cefas; Local Authorities; Industry discharge records; Water Companies |
| Tourism and nature watching | Number of participants (number per yr); number of facilities (number visitors per facility/yr); amount of time spent participating (hours/days) | Office for National Statistics; UK Centre for Economic & Environmental Development (CEED); Great Britain Tourism Survey; OBIS SEAMAP; RSPB statistics; Royal Yachting Association |
| Spiritual and cultural well-being | Sites with cultural heritage/well-being (usage rates by people, degree of importance); sites with spiritual and/or religious significance/well-being (number of people who attach significance, degree of significance attached) | Office for National Statistics; Economic and Social Data Service (ESDS) |

| | | |
|---|---|---|
| Aesthetic benefits | Number and/or area of marine features of given stated appreciation; length of heritage coast (km) | Office for National Statistics; Economic and Social Data Service (ESDS) |
| Education, research | Field trips (number and number of people involved); classes (numbers and number of people involved); scientific studies (number of research papers, subscriptions, library borrowing, on-line downloads); books (number, print run, library usage, e-book downloads); other publications including newspaper articles (circulation including on-line accessing); works of art (number of works, number of people viewing work) | Office for National Statistics; UK Directory of the Marine Observing Systems (UKDMOS); School and University Reports; Charting Progress 2 |
| Health benefits | % cover of coastal and marine environments; % cover of designated coastal and marine spaces (SACs, SPAs, EMS, MPAs, MCZs); time spent in the coastal/marine environment (hours); participation in particular activities in the coastal/marine environment (type and duration) | Natural England's 'Monitor of Engagement with the Natural Environment' (MENE) survey data |

and marine systems and in the case of performance indicators will require a set of associated targets to be established. Some indicators have strong links to management (e.g. the quantity and quality of fish and shellfish, amount of carbon sequestered), while in the case of others it can be argued that these reflect or are linked to important relationships with ecosystem processes and services (such as depth (m), volume ($m^3$), area of surface (ha), and tidal range (m)). For example, water depth (m) as an indicator has relevance to and may be a surrogate for the bathymetry and topography of a site, which in turn relate to the hydrographic regime. Water depth may impact in various ways on recreational activities, sea defence, coastal erosion, fisheries and aquaculture, and other services, but the nature of its impacts are also dependent on other marine components, processes and services.

Examples of UK data sources have been identified for each category of ecosystem service indicator (Table 5.1). These sources offer good spatial coverage at the UK-level, and contain both observed and modelled data. Such data tend to be available for long time series and regular updating can be anticipated in most instances. However, there can be a lack of consistency in the selection of spatial accounting units, and in the frequency of reporting, between data series. Where assessments are based on temporal or spatial extrapolation of evidence typically it becomes less meaningful due to the assumptions that are imposed in extending their coverage. Issues surrounding the quality of data, the complications associated with indicators capturing provision and utilisation (for example in natural waste management and natural hazard protection), areas of uncertainty and use of data are discussed in Chap. 8, although the particular focus of this chapter is on the social and economic data. The list of data sources presented in this chapter is not exhaustive but is used to suggest the range of national sources currently available. Where UK-level data sets have not been identified, evidence may be available from site-specific published and grey literature (as reported in Table 5.1 and examples are provided in the case studies below).

Although beyond the scope of this chapter, there may be a need for a further set of indicators which show the emergent properties of ecosystems (e.g. Basset et al. 2013). The emergent properties are linked to the fundamental processes of the system which provide the conditions for the delivery of ecosystem services. These properties include resistance, defined as the ability of a system to withstand the pressure caused by a potential stressor before it changes, and also resilience which is defined as the ability of the system to recover after the addition of a stressor (see Elliott et al. 2007). Indicators may relate to the structure of the system (i.e. attributes at one time) or functioning (i.e. related to rate processes). Indicators measuring the vulnerability of the system would allow policy makers to prioritise areas for action (e.g. Pethick and Crooks 2000). Similarly, indicators may be required for the ecological or socio-economic carrying capacity of the system and its ability to support ecological components or the human activities present (Elliott et al. 2007).

Indicators may also be used to assess whether a system is 'fit-for-purpose' or healthy. While there is still debate regarding the concept of 'health' in the marine environment (Tett et al. 2013), stakeholders and policymakers are aware of what constitutes at least aspects of a natural or healthy system (Mee et al. 2008), and indicators on these aspects are required by managers driven by legislation. In a European context, the implementation of both the WFD and the MSFD rely heavily on defining indicators against which respectively Good Ecological Status (GES) and Good Environmental Status (GEnS) are judged (Hering et al. 2010; Borja et al. 2010). The MSFD relates more to the functioning of the marine system as it focuses on 11 descriptors as opposed to the structural approach of the WFD. There is currently an on-going debate on the set of indicators identified for the MSFD by the relevant Descriptor Task Teams for assessing GEnS (Borja et al. 2013) although, as yet, there are no firm directions on the way in which the indicators may be combined. The indicators for the MSFD are currently more open to interpretation by individual member states, than those identified in Table 5.1, which is considered necessary given the variation in the characteristics of the marine environment of the regional seas across Europe. It is notable that Borja et al. (2013) argue the requirement for indicators of GEnS to incorporate ecosystem services and goods/benefits given the need for the marine environment to not only protect and enhance the nature conservation features but also to deliver ecosystem services and societal benefits.

Issues surrounding indicators (Table 5.1) are illustrated further with reference to fisheries and aquaculture and carbon sequestration and storage. These two examples are used to demonstrate how multiple indicators may be necessary to reflect the complexity of coastal and marine systems associated with even single services and to detect change over time in their provision. However, it is unlikely that indicators for all elements of the ecosystem services framework will be used in such a case; hence we identify indicators pertaining to a sub-set of the framework in each case.

Figure 5.1 focuses on ecosystem indicators associated with coastal and marine fisheries and aquaculture. It specifically relates to the supply of wild or farmed food for human consumption as a good/benefit. Since complementary (man-made) capital is required to obtain the goods/benefits from ecosystem services, in the interpretation of indicators of wild or farmed food, knowledge is required of the wide range of complementary capital employed since that harvested for human consumption is dependent on the level of effort exerted (e.g. number of days at sea and number of pots/trawls). For clarity, the figure omits indicators of complementary capital. The final ecosystem service 'coastal and marine biota' should obviously be highlighted as being of particular importance to food for human consumption. While these final ecosystem service indicators are typically species-specific, it is recognised that indicators of the four intermediate services linked to these are likely to be broadly similar for all species and it is self-evident that no single indicator will satisfy all requirements. These intermediate services relate to 'primary production', 'larval and gamete supply', 'nutrient cycling', and 'formation of species habitat'. An interesting case is the relationship between indicators of fish stocks and seabird populations (e.g. Kittiwakes numbers) which compete for the same food resource (e.g. indicated by sand-eel numbers) and implies that careful interpretation is required regarding the causes and effects of change for example in the trophic system. In the case of fisheries and aquaculture much of the complexity of the coastal and marine system of direct relevance to this service can be reflected by indicators associated

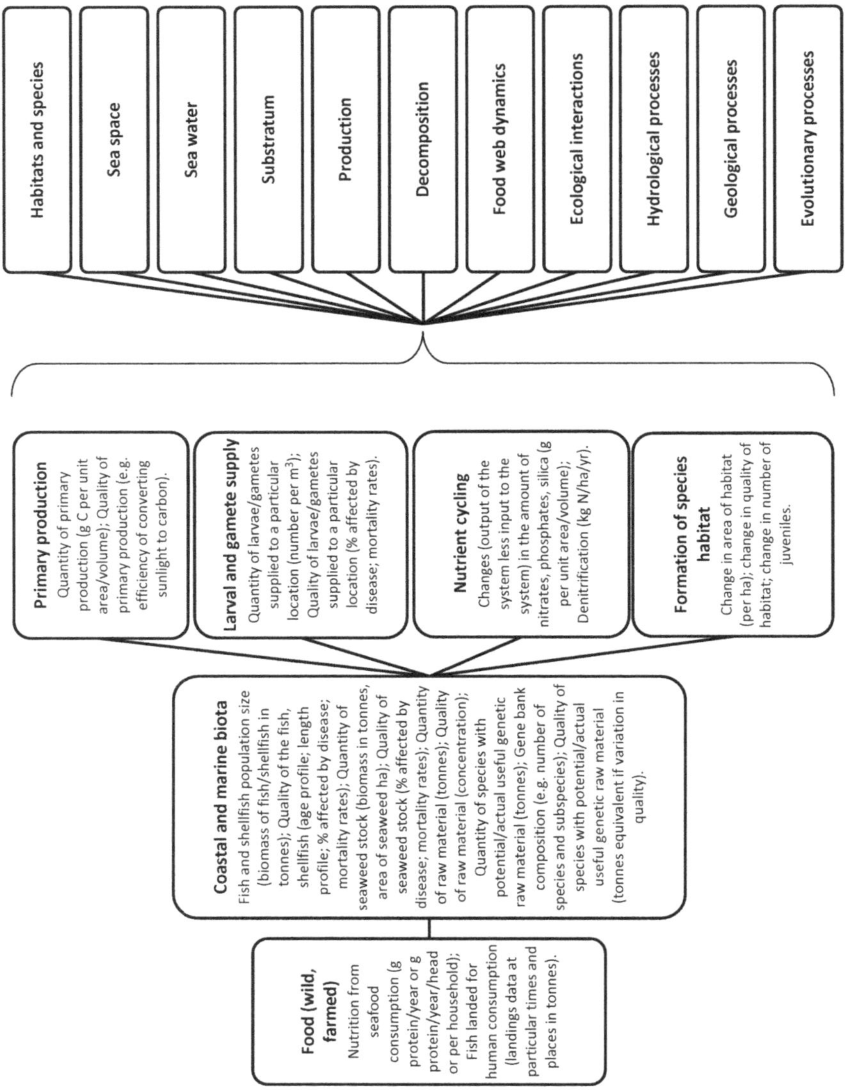


**Fig. 5.1** Ecosystem service indicators associated with coastal and marine fisheries and aquaculture

with the first three columns given in Fig. 5.1. While the importance of the underlying coastal and marine ecosystem components and processes are recognised in the fourth column of the figure, and indicators are listed in Table 5.1, it is likely that indicators of these attributes of the marine environment provide a less clear evidence base given uncertainties associated with cause and effect relationships, and therefore the ecological bottom-up approach would not be advocated. Also, given the openness of the coastal and marine system, the delivery of an ecosystem service in one area may be dependent on other services in areas outside the system-in-focus; for example, the size of the fishable stock in a UK coastal area may depend on feeding and nursery grounds in other areas. This is particularly the case in ecosystems with a very high degree of connectivity such as estuaries and coasts (Elliott and Whitfield 2011) in which fundamental processes and the well-being of biological elements such as juvenile fishes (a nursery function) or wading and sea birds (cultural services) depend on features occurring outside of that system. Similarly, the response of a system to nutrients depends on the nutrient fluxes elsewhere. Ecosystem service indicators related to fisheries are further discussed in Sect. 5.4.2 in relation to the role of managed realignment sites in providing nursery habitats for commercial fish species.

Figure 5.2 focuses on appropriate indicators for changes in coastal and marine carbon sequestration and storage, identified within the UK NEAFO ecosystem services framework as an intermediate ecosystem service. To reflect the complexity of the coastal and marine system associated with this service, a wide set of indicators is required of the system's components and processes, along with indicators of climate regulation (a final service) and healthy climate (a good/benefit). Again, indicators have not been given in Fig. 5.2 for the components and processes, however a full list of indicators is provided in Table 5.1. Focusing on the components and processes, both biological indicators (relating to habitats and species, production, decomposition, food web dynamics, ecological interactions) and physical indicators (relating to substratum, hydrological processes and geological processes) create the capacity for carbon sequestration with the outcome dependent on the ecological and physico-chemical characteristics of the specific site. Relevant indicators at this level include abundance (no.), cover (%), or biomass (g, kg) of particular habitats and species, changes in abundance/biomass over time, area (ha) and/or depth (m) of substratum by type (mud, sand, gravel etc.), sediment accumulation rates (cm per year), channel depth (m), etc. The complexity of coastal and marine sites combined with availability of data may lead to attention being focussed on indicators of 'climate regulation' and/or of 'healthy climate'. However, in particular the good/benefit in this context (healthy climate) has different spatial dimensions to the components and processes as reflected in the definition and formulation of the associated indicators. This again raises issues about the boundary conditions of the system-in-focus and makes it problematic to attribute any change in an indicator of healthy climate to the state of a specific site, an activity or a management initiative (Atkins et al. 2011). Data requirements for applying indicators for carbon sequestration and storage are discussed in the managed realignment case study below (Sect. 5.4.2), and in Chaps. 10 and 11.

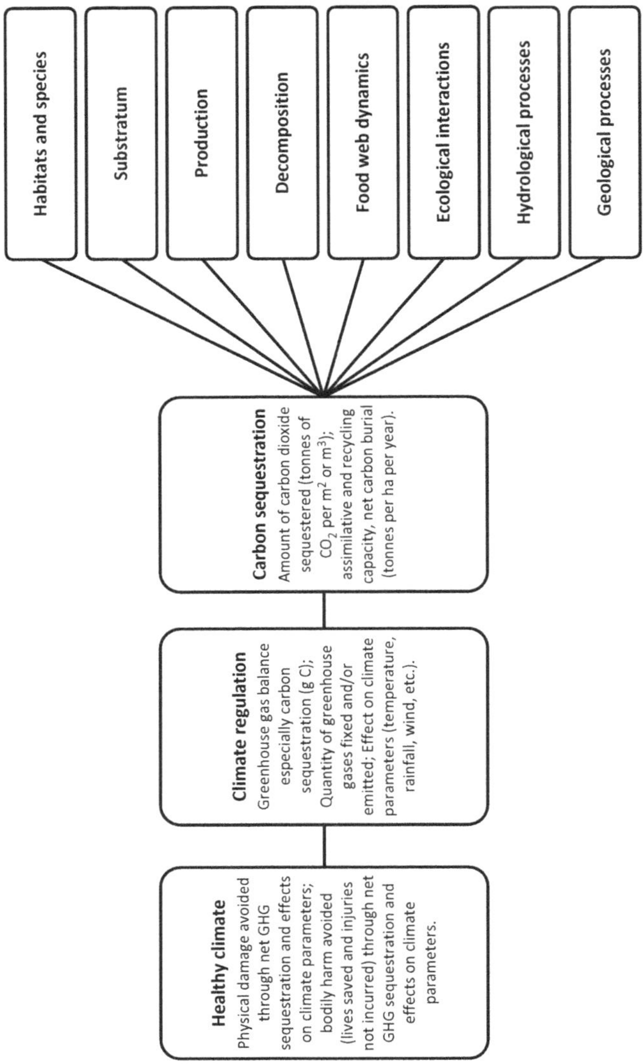


**Fig. 5.2** Ecosystem service indicators relating to carbon sequestration and storage in the coastal and marine environment

Although our discussion of ecosystem service indicators has been in the context of the UK NEAFO framework and its ecosystem service categories, indicators identified here could, for example, be applied to coastal and marine biodiversity which is not considered here to be an ecosystem service but is nevertheless of interest. In this case it is unlikely that indicators from one element of the ecosystem services framework alone will suffice to capture state, trend and/or behaviour. For example, the state of marine biodiversity, which can be a site-specific attribute and one that underpins the provision of marine ecosystem services, would be most appropriately captured using a wide range of indicators associated with those marine components relating to 'habitats and species'.

## 5.4 Case Studies

### *5.4.1 Marine Protected Areas*

In the context of Marine Protected Areas (MPAs), Potts et al. (2014) recognised the importance of ecosystem service provision by existing and proposed UK MPAs, for example the Lundy No Take Zone (NTZ) and Skomer Marine Nature Reserve (MNR). Linking conservation management, in the form of MPAs, and human well-being is further discussed in Chap. 9. The requirement for a suite of ecosystem service indicators to support the selection, monitoring and evaluation of associated marine management measures is apparent. The data for such indicators are likely to be best drawn from site-specific published and unpublished sources given the finer resolution which would be required at a more localised level. Some examples of the site-specific data sources are provided by a partial review of evidence relevant to ecosystem service indicators for the two MPA cases.

The Lundy NTZ was designated in 2003 in order to protect marine wildlife while improving local fish stocks, and is located to the east of Lundy Island, within the wider Lundy Marine Conservation Zone (MCZ). Evidence for the site can be found in annual Lundy Field Society reports (dating back to 1947) and in more recent studies which focus specifically on the impacts of the NTZ (Hoskin et al. 2009; Hoskin et al. 2011; Wootton et al. 2012; Coleman et al. 2013). Evidence suggests that there have been changes to the provision of a number of ecosystem services since its designation, and these have included improvements in local shellfish stocks (a final ecosystem service), potential spill-over effects to the local fishery (a good/benefit), and improvements to local tourism/nature watching (a good/benefit) reflected in on-site recreational diving. Examples of the available ecosystem service indicator evidence include:

- Hoskin et al. (2009) on the first 5 years of the NTZ suggested that for local shellfish stocks there was a change in the size profile of the population with a 5 % increase in the size of European lobster (*Homarus gammarus*) and a 427 % increase in abundance of European lobster within the Lundy NTZ (see Fig. 5.1).

It is suggested that an observed 97 % increase in the abundance of undersized lobsters within the NTZ and 124 % increase in its abundance in waters adjacent to the NTZ boundary provides potential evidence of a spill-over effect to the local lobster fishery – an assessment from this extended area of lobster landings for human consumption would be required to confirm this suggestion. These findings are supported by Wootton et al. (2012) who used similar methods to demonstrate positive effects of the Lundy NTZ on increased lobster abundance and size within the NTZ.

- Wootton et al. (2012) demonstrated the apparent negative effects of the NTZ including increased injury and shell disease. Their study raised concerns about the impact that greater population densities has on disease outbreaks (% with disease), with evidence suggesting that high severity shell disease in the Lundy NTZ was significantly associated with injury, for example injured male lobsters within the NTZ were over three times more likely to possess the high severity form of shell disease.
- The wider Lundy MCZ attracts a large number of recreational divers; it was estimated that 1,370 recreational diver days (1 person diving for 1 day) occur at Lundy each year, around 60 % of which occur within the NTZ (equating to 820 diver days) (MCZ Project 2012). An improvement in the condition of site features, including any associated increase in abundance and diversity of species, could improve the quality of diving at the site and impact on well-being through increased diving participation rates.

Skomer MNR, which includes the waters around Skomer Island, Middleholm and parts of the Marloes Peninsula in South Pembrokeshire is currently the only MNR in Wales; it was designated in 1990 previously having been a voluntary MNR, and is managed by Natural Resources Wales. Evidence to support the use of a suite of indicators can be found in the Skomer MNR annual reports (providing some data series back to 1987), various reports produced by the Countryside Council for Wales (now Natural Resources Wales), Joint Nature Conservation Committee, and similar agencies. Examples of such evidence include:

- The Countryside Council for Wales reports that the restriction of mobile fishing gears within the Skomer MNR has increased the abundance of the local King scallop (*Pecten maximus*) population 'at least four fold and perhaps more than eight fold' over the first 20 years of the MNR's designation (CCW Press Release 20 April 2010).
- Taylor et al. (2012) reported the number of breeding birds (counting Apparently Occupied Nests, AON) and breeding success of Black-legged Kittiwakes (*Rissa tridactyla*) on Skomer Island between 1989 and 2012. In 2012, the breeding numbers of Black-legged Kittiwake totalled 1,594 AON, which was a 13.23 % decrease on 2011 breeding numbers and a 30.15 % decline over the last 5 years (2007–2012). The success of fledging Black-legged Kittiwakes is a recognised OSPAR Ecological Quality Objective and was reported for 2012 at three sites on Skomer Island (S. Stream, High Cliff, and The Wick), encompassing 37 % of the total breeding population. The mean success of fledging Black-legged Kittiwakes

was reported as 0.32 per AON for 2012, which is below the 23-year mean of 0.64 per AON since monitoring began.

- Monitoring of Grey seal nursery areas in 2012 indicated that the total pup numbers for the MNR reached 310 which is the highest total ever recorded – pup survival was 76 %, which is slightly below the average for the last 10 years (Lock et al. 2013).
- The site is also recognised as providing services associated with tourism/wildlife watching with numbers of participants reported annually. For example, 1,008 diver days (with the Lucy wreck located within the MNR a popular dive site), 380 recreational craft visits and 483 anglers (192 shore and 291 boat anglers) were recorded within the Skomer MNR for 2012 (Lock et al. 2013). The available time series data reveals changes in participation rates over time.

These two marine sites have been of conservation interest for many years, and have been subject to extensive compulsory monitoring of the features of conservation interest along with other scientific investigation. Hence, their evidence base will not be typical of that available for other UK marine sites if those sites do not have equivalent designations and therefore no legal requirement for such monitoring.

### *5.4.2 Managed Realignment*

The importance of saltmarsh habitat in the UK has been recognised within the literature for the delivery of a number of ecosystem services, including sea defence, prevention of coastal erosion, formation of species habitat for birds, fish and invertebrates, carbon sequestration and tourism/nature watching (Everard 2009; Fonseca 2009; Luisetti et al. 2011; Burdon et al. 2011). The restoration of saltmarsh habitat, through the implementation of managed realignment (MR) at appropriate sites within the UK may provide a wide range of ecosystem services (Luisetti et al. 2011). Mander et al. (2013) showed the potential for MR sites as wading bird feeding areas but emphasised that while they provide additional feeding time, their prey carrying capacity (and hence creation of certain ecosystem services) may not be the same as natural areas.

In order to assess change in ecosystem service provision, a suite of ecosystem service indicators may be required.

- King and Lester (1995) compared the sea defence capacity of man-made sea defence structures with and without a saltmarsh buffer. Using the width of saltmarsh (in m) as an indicator (see Table 5.1), their study deduced that as the width of vegetation decreases, the height of the man-made sea wall would need to increase in an almost linear relationship, with related cost implications. For example, at a site with an 80 m width of saltmarsh habitat a 3 m high sea wall would be required for sea defence provision, whereas if the saltmarsh habitat was removed, a 12 m high sea wall would be required to provide the same level of sea defence.

- Möller et al. (2002) used energy dissipation capacity as an indicator for natural hazard protection (a final ecosystem service) which is of relevance to both sea defence and prevention of coastal erosion (see Table 5.1), and reported energy dissipation rates of 89 % over saltmarsh as opposed to 29 % over bare sand flats.
- Several studies have examined the potential for saltmarsh to act as a nursery and/or feeding area for fish species. For example, Fonseca (2009) used the abundance of juvenile sea bass (*Dicentrarchus labrax*) per size class within three MR sites in the Blackwater Estuary, in combination with an estimate of the survival rate to minimum commercial landing size per size class, as an indicator of the potential contribution to local fish stocks. Findings show that the sampled MR sites have the potential to contribute 1.65 kg of juvenile bass per hectare of saltmarsh (mean value) surviving to minimum landing size (36 cm) after 4 or 5 years. This example is further discussed in Chap. 11.
- Saltmarsh is recognised as providing an important carbon sequestration service within MR sites (Fig. 5.2), for example Luisetti et al. (2011) reported net carbon burial values (in tonnes per ha per year) of 0.266 and 3.347 for sedimentation rates of 1.5 mm and 6 mm respectively within the Blackwater Estuary. This example is further discussed in Chap. 10.
- MR sites provide potential for recreational activities, with the number of participants per activity being identified as a suitable indicator of the state of this good/benefit (Table 5.1). For example, it was reported that the most popular uses of the Paull Holme Strays MR site in the Humber Estuary included walking/running (61 % of respondents), enjoyment of the site/fresh air (59 % of respondents), dog walking (41 % of respondents), bird/nature watching (37 % of respondents) and fishing (10 % of respondents) (n = 117, Environment Agency 2007). Changes in the relative importance of these activities as the site develops would provide a useful indicator of the behaviour and trend of the provision of these services over time.

If managed realignment is proposed within an estuary or coastal environment protected for its nature conservation value, there is a legal requirement for each realignment site to have its own monitoring and management strategy related to its objectives (e.g. habitat compensation linked to a legally binding offset agreement). Indicator use in this setting is therefore relatively well established for both biotic features (although often limited to monitoring of bird, vegetation, and benthic communities) and abiotic features (often limited to accretion monitoring). However, the difficulty of interpreting such monitoring evidence against the background of a highly dynamic, naturally changing system is noted. Changes in the provision of ecosystem services following managed realignment are further discussed in Chap. 11.

## 5.5 Conclusion

This chapter has identified a practicable set of indicators for the coastal and marine environment, grounded within an ecosystem services framework, which capture key elements of the DPSIR State changes and Impacts. Since the framework distinguishes between coastal and marine processes and components, intermediate ecosystem services, final ecosystem services and goods/benefits, a basis is provided for establishing a suite of indicators that can reflect in a systematic way the interrelationships and causality, thereby being consistent and comprehensive in their coverage of the natural, economic and social dimensions of coastal and marine systems. Attention is also drawn to the need for indicators to be SMART. In these ways, the indicators are linked to and can be part of a decision support system for adaptive management.

Increasingly ecosystem services are being incorporated into coastal and marine policy and management to recognise the impact of environment change on human well-being and, in the case of habitat restoration to account for historical degradation, for assessing our ability to recreate the original services. It is important that with the employment of indicators in such uses, effort is invested in defining and formulating the indicators, which aids communication between managers, policy-makers and stakeholders, but it is imperative also to evaluate their relevance and the precision of estimation. If indicators are used to determine management actions, for example the desire to obtain a certain amount of a service or to prevent a decline in service provision beyond a given threshold, then if the desired outcome is not met then this may result in legal challenges. Such legal challenges are likely to focus on the quality of the science presented and the adequacy of the indicator applied, including attention on any uncertainties surrounding the choice of indicator, its SMART qualities, and in its population with data. Borja and Elliott (2013) caution that financial restrictions on monitoring and data collection will increase these uncertainties.

## References

Atkins, J. P., Burdon, D., Elliott, M., & Gregory, A. J. (2011). Management of the marine environment: Integrating ecosystem services and societal benefits with the DPSIR framework in a systems approach. *Marine Pollution Bulletin, 62*, 215–226.

Aubry, A., & Elliott, M. (2006). The use of environmental integrative indicators to assess seabed disturbance in estuaries and coasts: Application to the Humber Estuary, UK. *Marine Pollution Bulletin, 53*, 175–185.

Basset, A., Elliott, M., West, R. J., & Wilson, J. G. (2013). Estuarine and lagoon biodiversity and their natural goods and services. *Estuarine, Coastal and Shelf Science, 132*, 1–4.

Böhnke-Henrichs, A., de Groot, R., Baulcomb, C., Hussain, S., Koss, R., & Fofana, A. (2013). Typology and indicators of ecosystem services for marine spatial planning and management. *Journal of Environmental Management, 130*, 135–145.

Borja, A., & Elliott, M. (2013). Marine monitoring during an economic crisis: The cure is worse than the disease. *Marine Pollution Bulletin, 68*, 1–3.

Borja, A., Elliott, M., Carstensen, J., Heiskanen, A.-S., & van de Bund, W. (2010). Marine management – Towards an integrated implementation of the European marine strategy framework and the water framework directives. *Marine Pollution Bulletin, 60*, 2175–2186.

Borja, A., Elliott, M., Andersen, J. H., Cardoso, A. C., Carstensen, J., Ferreira, J. G., Heiskanen, A.-S., Marques, J. C., Neto, J., Teixeira, H., Uusitalo, L., Uyarra, M. C., & Zampoukas, N. (2013). Good environmental status of marine ecosystems: What is it and how do we know when we have attained it? *Marine Pollution Bulletin, 76*, 16–27.

Burdon, D., Atkins, J. P., & Bhatia, N. (2011). *Ecosystem services and societal benefits provided by estuarine saltmarsh*. Report No. SBB338-F1-2011. Institute of Estuarine & Coastal Studies, University of Hull, Hull, HU6 7RX, UK.

Coleman, R. A., Hoskin, M. G., von Carlshausen, E., & Davies, C. M. (2013). Using a no-take zone to assess the impacts of fishing: Sessile epifauna appear insensitive to environmental disturbances from commercial potting. *Journal of Experimental Marine Biology and Ecology, 440*, 100–107.

Dale, V. H., & Beyeler, S. C. (2001). Challenges in the development and use of ecological indicators. *Ecological Indicators, 1*, 3–10.

Dauvin, J. C., Fisson, C., Garnier, J., Lafite, R., Ruellet, T., Billen, G., Deloffre, J., & Verney, R. (2008). A report card and quality indicators for the seine estuary: From scientific approach to operational tool. *Marine Pollution Bulletin, 57*, 187–201.

De Groot, R. S., Alkemade, R., Braat, L., Hein, L., & Willemen, L. (2010a). Challenges in integrating the concept of ecosystem services and values in landscape planning, management and decision making. *Journal of Ecological Complexity, 7*, 260–272.

De Groot, R. S., Fisher, B., Christie, M., Aronson, J., Braat, L., Haines-Young, R., Gowdy, J., Maltby, E., Neuville, A., Polasky, S., Portela, R., & Ring, I. (2010b). Integrating the ecological and economic dimensions in biodiversity and ecosystem service valuation. In P. Kumar (Ed.), *The economics of ecosystems and biodiversity: Ecological and economic foundations* (pp. 9–40). London/Washington, DC: Earthscan.

Doran, G. T. (1981). There's a S.M.A.R.T. way to write management's goals and objectives. *Management Review, 70*, 35–36.

Elliott, M. (2011). Marine science and management means tackling exogenic unmanaged pressures and endogenic managed pressures – a numbered guide. *Marine Pollution Bulletin, 62*, 651–655.

Elliott, M., & Whitfield, A. (2011). Challenging paradigms in estuarine ecology and management. *Estuarine, Coastal and Shelf Science, 94*, 306–314.

Elliott, M., Burdon, D., Hemingway, K. L., & Apitz, S. (2007). Estuarine, coastal and marine ecosystem restoration: Confusing management and science – A revision of concepts. *Estuarine, Coastal and Shelf Science, 74*, 349–366.

Ellis, J.R., Milligan, S.P., Readdy, L., Taylor, N., & Brown, M.J. (2012). Spawning and nursery grounds of selected fish species in UK waters. *Science Ser Technology Report, Cefas Lowestoft, 147*, 56 pp.

Environment Agency. (2007). *Paull Holme Strays resident survey*. Research findings prepared by BRAHM for the Environment Agency, UK.

Everard, M. (2009). *Ecosystem services case studies*. Environment Agency Report. Rio House, Waterside Drive, Aztec West, Almondsbury, Bristol, BS32 4UD.

Fonseca, L. (2009). *Fish utilisation of managed realignment areas and saltmarshes in the Blackwater Estuary, Essex, S.E. England*. PhD Thesis, Queen Mary University of London.

Gabrielsen, P., & Bosch, P. (2003). *Environmental indicators: Typology and use in reporting*. European Environment Agency Internal Working Paper, August 2003.

Gibbs, M. T. (2012). The conundrum of marine environmental indicators: The argument for new marine sensor technology. *Ecological Indicators, 13*, 338–341.

Gray, J. S., & Elliott, M. (2009). *Ecology of marine sediments: Science to management*. Oxford: Oxford University Press.

Hattam, C., Atkins, J. P., Beaumont, N., Borger, T., Bohnke-Henrichs, A., Burdon, D., De Groot, R., Hoefnagel, E., Nunes, P., Piwowarczyk, J., Sergio, S., & Austen, M. (2015). Marine ecosystem services: Linking indicators to their classification. *Ecological Indicators, 49*, 61–75.

Hering, D., Borja, A., Carstensen, J., Carvalho, L., Elliott, M., Feld, C. K., Heiskanen, A. S., Johnson, R. K., Moe, J., Pont, D., Solheim, A. L., & van de Bund, W. (2010). The European water framework directive at the age of 10: A critical review of the achievements with recommendations for the future. *Science of the Total Environment, 408*, 4007–4019.

Hoskin, M. G., Coleman, R. A., & von Carlshausen, L. (2009). *Ecological effects of the lundy No-take zone: The first five years (2003–2007)*. Report to Natural England, DEFRA and WWF-UK.

Hoskin, M. G., Coleman, R. A., von Carlshausen, L., & David, C. (2011). Variable population responses by large decapods crustaceans to the establishment of a temperate marine no-take zone. *Canadian Journal of Fisheries and Aquatic Sciences, 68*, 185–200.

Kandziora, M., Burkhard, B., & Müller, F. (2012). Interactions of ecosystem properties, ecosystem integrity and ecosystem service indicators – a theoretical matrix exercise. *Ecological Indicators, 28*, 54–78.

King, S. E., & Lester, J. N. (1995). The value of saltmarsh as a sea defence. *Marine Pollution Bulletin, 30*, 180–189.

Kumar, P. (2010). *The economics of ecosystems and biodiversity: Ecological and economic foundations*. London/Washington, DC: Earthscan.

Liquete, C., Piroddi, C., Drakou, E. G., Gurney, L., Katsanevakis, S., Charef, A., & Egoh, B. (2013). Current status and future prospects for the assessment of marine and coastal ecosystem services: A systematic review. *PLoS ONE, 8*(7), e67737.

Lock, K., Newman, P., Burton, M., & Jones, J. (2013). Skomer Marine nature reserve annual report 2012, countryside council for Wales Regional Report CCW/WW/13/4, March 2013.

Luisetti, T., Turner, R. K., Bateman, I. J., Morse-Jones, S., Adams, C., & Fonseca, L. (2011). Coastal and marine ecosystem services valuation for policy and management: Managed realignment case studies in England. *Ocean and Coastal Management, 54*, 212–224.

MEA. (2005). *Millennium ecosystem assessment – Ecosystems and human wellbeing biodiversity synthesis*. Washington, DC: Island Press.

Mander, L., Marie-Orleach, L., & Elliott, M. (2013). The value of wader foraging behaviour study to assess the success of restored intertidal areas. *Estuarine, Coastal and Shelf Science, 131*, 1–5.

McLusky, D. S., & Elliott, M. (2004). *The estuarine ecosystem; ecology, threats and management* (3rd ed.). Oxford: Oxford University Press.

MCZ Project. (2012). Finding sanctuary, Irish Seas conservation zones, net gain and balanced seas – Impact assessment materials in support of the regional marine conservation zone projects' recommendations, 13 July 2012.

Mee, L. D., Jefferson, R. L., Laffoley, D. d'. A., & Elliott, M. (2008). How good is good? Human values and Europe's proposed marine strategy directive. *Marine Pollution Bulletin, 56*, 187–204.

Möller, I., Spencer, T., & Rawson, J. (2002). Spatial and temporal variability of wave attenuation over a UK East-coast saltmarsh. In *Proceedings of the 38th international conference on coastal engineering, Cardiff, July 2002*. London: Thomas Telford Publishing.

Niemeijer, D., & de Groot, R. S. (2008). A conceptual framework for selecting environmental indicator sets. *Ecological Indicators, 8*, 14–25.

Pethick, J. S., & Crooks, S. (2000). Development of a coastal vulnerability index: A geomorphological perspective. *Environmental Conservation, 27*, 359–367.

Potts, T., Burdon, D., Jackson, E., Atkins, J. P., Saunders, J., Hastings, E., & Langmead, O. (2014). Do marine protected areas deliver flows of ecosystem services to support human welfare? *Marine Policy, 44*, 139–148.

Smeets, S., & Weterings, R. (1999). *Environmental indicators: Typology and overview*. EEA Technical report No. 25, Copenhagen.

Taylor, C. J., Boyle, D., Perrins, C. M., & Kipling, R. (2012). *Seabird monitoring on Skomer Island in 2012*. JNCC Report.

Tett, P., Gowen, R., Painting, S., Elliott, M., Foster, R., Mills, D., Bresnan, E., Capuzzo, E., Fernandes, T., Foden, J., Geider, R., Gilpin, L., Huxham, M., McQuatters-Gollop, A., Malcolm, S., Saux-Picart, S., Platt, T., Racault, M.-F., Sathyendranath, S., van den Molen, J., & Wilkinson, M. (2013). Framework for understanding marine ecosystem health. *Marine Ecology: Progress Series, 494*, 1–27.

TIDE. (2013). *Management of Estuaries: The need to understand nature and society*. TIDE summary report. The North Sea Region Programme (2007–2013) Interreg IVB project. Hamburg Port Authority (Lead Partner), Germany.
Turner, R. K., Mee, L., Elliott, M., Burdon, D., Atkins, J., Potts, T., Jickells, T., Bee, E., & Beaumont, N. (2013). *Coastal zone ecosystem services: From science to values and decision making, a conceptual framework*. Valuing nature network report, January 2013. University of East Anglia, Norwich.
UK National Ecosystem Assessment. (2011). *The UK National Ecosystem Assessment: Synthesis of the key Findings*. Cambridge: UNEP-WCMC.
UNEP-WCMC. (2009). *Ecosystem service indicators: Gaps, opportunities, and next steps*. Report of the expert workshop on ecosystem service indicators, Cambridge, 22-23/09/09.
UNEP-WCMC. (2011). *Developing ecosystem service indicators: Experiences and lessons learned from sub-global assessments and other initiatives*. Secretariat of the Convention on Biological Diversity, Montréal, Canada. Technical Series No. 58.
Van Den Belt, M., & Costanza, R. (2011). Ecological economics of estuaries and coasts. In D. S. McLusky & E. Wolanski (Eds.), *Treatise on estuarine & coastal science. A 12 volume set*. London: Academic.
Wootton, E. C., Woolmer, A. P., Vogan, C. L., Pope, E. C., & Hamilton, K. M. (2012). Increased disease calls for a cost-benefits review of marine reserves. *PLoS ONE, 7*(12), e51615.

# Chapter 6
# Valuation of Coastal and Marine Ecosystem Services: A Literature Review

**M. Schaafsma and R.K. Turner**

## 6.1 Introduction

Valuation can be used to support the step of economic (and social) appraisal and valuation of options in the adaptive coastal management approach. This chapter aims to assess the availability of primary valuation studies providing economic value estimates for 'goods and benefits' generated from coastal and marine ecosystems. This overview reveals the main gaps in the literature with respect to primary (monetary) valuation studies addressing coastal and marine habitats and specific ecosystem services, globally and in particular for Great Britain (GB). We assess the extent to which monetary value estimates of the ecosystem goods and benefits and habitat types that are most important in GB are available from the literature.

The assessment and valuation of ecosystem stock and flow situations is not straightforward and some goods and benefits cannot be meaningfully valued in monetary terms (those related to cultural services in particular) (see Chap. 4). This chapter only covers monetary value estimates based on economic valuation methods to derive marginal *economic* values of changes in the flow of goods and benefits over time. Such marginal values can be used in support of decision making on trade off choices. A number of reports (Posford Duvivier Environment 1996; Pugh and Skinner 2002; Pugh 2008; Saunders et al. 2010; UKMMAS 2010) review the *financial* values (e.g. in terms of gross value added (GVA) to the UK economy) of marine-dependent industries, including fisheries and tourism.

---

M. Schaafsma (✉)
Geography and Environment and Centre for Biological Sciences, University of Southampton, University Road, Southampton SO17 1BJ, UK
e-mail: M.Schaafsma@soton.ac.uk

R.K. Turner
School of Environmental Sciences, University of East Anglia, Norwich, UK



R.K. Turner, M. Schaafsma (eds.), *Coastal Zones Ecosystem Services*, Studies in Ecological Economics 9, DOI 10.1007/978-3-319-17214-9_6

## 6.2 Methodology

### 6.2.1 Scope

The literature review is structured around a particular set of habitats and ecosystem goods and benefits. Many ecosystem services assessments aim to map values onto habitats or 'land cover- land use' maps. There are, however, multiple habitat classifications for coastal and marine areas, which are, for example, habitat, depth, salinity or sediment based. For the purpose of this valuation literature review, we aim to map the valuation studies onto the six coastal margin and six marine habitats identified in the UK NEA (2011), two tropical habitats (coral reefs and mangroves) and two complex habitats (estuaries – including fjords and bays – and intertidal wetlands). The latter are included because habitats covered by valuation studies are sometimes less precise or more pragmatically defined. For example, estuaries may encompass different intertidal (e.g. mud flats) and shallow subtidal areas (e.g. seagrass beds, kelp forest), as well as coastal margins (e.g. salt marshes) (Moss 2008). Because of this habitat complexity, valuation studies often do not or cannot assign ecosystem goods and benefits to specific habitat types within an estuary and broadly label the study area as an estuary. Similarly, valuation studies report to provide values for '(intertidal) wetlands', which may encompass other habitats, such as marshes and mudflats. Other valuation studies do not provide sufficient detail about the study area to assign values to either intertidal or subtidal areas.

Valuation studies included in the 'coastal shelf' category may include different coastal and marine habitat types, depending on the study area. Where possible, we allocated these studies to specific habitat types, but when this was impossible the study was included in the (therefore broad) coastal shelf category. In addition, we assigned economic values to the coastal shelf if the fisheries pertained to political Exclusive economic zones (EEZ).

This literature overview focuses on the following goods and benefit categories (see Fig. 2.6 in Chap. 2): products (food, bait and fish feed, fertiliser, etc.); 'healthy' climate; prevention of coastal erosion; sea defence; tourism and nature watching; spiritual and cultural wellbeing; aesthetic values; and education and research. The literature review excludes water purification services as sea water use for water supply is very limited (UKMMAS 2010), and benefits of improved water quality are captured in other categories, such as recreation and amenity, products or seascape values. Human health benefits are excluded as well, although they may partially be captured in recreational values (e.g. see Georgiou et al. 2000). Following the UK NEA ESF (see Chap. 2), we also excluded services that relate to abiotic components of the areas, and services with negative impacts related to off-shore wind farms, artificial reefs and pests or invasive species (see Chap. 12).

In valuation studies, the reported economic value may correspond to the benefits derived from a bundle of goods and benefits. This is especially the case for studies that aim at capturing values of tourism, nature watching and aesthetic benefits of meaningful seascapes. The value that people attach to certain wild species and natural habitats and seascapes values may reflect spiritual and cultural wellbeing, health benefits, and aesthetic values, and thus may contain an aspect of non-use (bequest, existence) values. In such cases, it is difficult to assign separate values to each of the ecosystem goods and benefits. For the purpose of this chapter, we have therefore created a category labelled "Spiritual and cultural well-being and aesthetic benefits of wild species and seascapes".

### 6.2.2 *Study Selection and Quality Criteria*

The overview covers primary valuation studies published since 2000 in academic journals and book chapters that have undergone peer-review. Papers published in grey literature (consultancy and non-governmental organisation reports, working papers) or before 2000 are excluded. Peer-revision is taken as a quality assessment of the analysis. Valuation estimates are subject to serious spatial and temporal bias constraints and in the latter context a period of more than a decade or so is a prudent limit. The selection process is based on web-searches in Science Direct and Google Scholar using the key-words 'ecosystem services', '(economic) valuation', 'coastal', 'marine', in various combinations. Primary studies referenced in the selected studies, available meta-analyses or other review papers (e.g. Beaumont et al. 2008, 2010) are included. Finally, a more targeted search on specific journals and authors is also performed to complete the list. The selection processes is limited to data available up to 1 May 2014.

From each selected study, we extracted information on the authors, year of publication, continent and country of the case study, valuation method, habitat type and ecosystem goods and services under consideration. To evaluate the completeness of the valuation evidence base for GB, we extract value estimates from GB-based studies and converted these to 2012 GBP prices and review the studies based on a number of criteria that qualify studies for benefit transfer purposes (Brouwer 2000). We focused on the adequacy of the data, soundness of economic methods, quality of the empirical techniques, and validity of the model or WTP function.

## 6.3 Results

### 6.3.1 *Descriptive Statistics*

The selection process resulted in 233 primary valuation studies, including 30 GB-based studies, published between 2000 and May 2014 in peer-reviewed academic journals and books. In addition, we identified nine relevant meta-analyses (Brander et al. 2006, 2007, 2012; Martín-López et al. 2008; Enjolras and Boisson 2010; Latinopoulos 2010; Londoño and Johnston 2012; Salem and Mercer 2012; Ghermandi and Nunes 2013).

There is no obvious positive trend in the number of publications over time. Stated preference (SP) methods, including contingent valuation (CV) and choice experiments (CEs), are used most frequently, mainly to assess recreational and biodiversity values, followed by travel cost (TC) assessments for recreational values and estimation of gross or net revenues to assess benefits of raw materials (mainly fishing). The majority of studies address case study areas in Europe, North-America (mostly USA) and Asia. A third of the European case studies are for the UK, but this may reflect an upward biased due to our focus on GB-based valuation evidence for the UK NEA FO. Table 6.1 provides an overview of the number of studies that provide economic values for each of the habitat – goods/benefits combinations.

Globally, 'tourism and nature watching' is the most frequently valued ecosystem benefit (67 % of the studies), followed by biodiversity and cultural values of habitats (33 %). This corresponds to the high numbers of SP and TC studies. Most of the tourism studies are for tropical coral reefs, beaches and coastal areas more broadly. There are very few valuation studies for ecosystem benefits related to prevention of coastal erosion (2 %), and education and research (1 %: Samonte-Tan et al. 2007, Cesar and van Beukering 2004). Surprisingly, only a small number of studies (4 %) are available for the carbon sequestration potential of coastal and marine habitats.

The distribution of studies across the different habitats shows that for sea cliffs and small islands (Chae et al. 2012), open oceans (i.e. beyond EEZ zones, Murillas-Maza et al. 2011) and cold water corals (Wattage et al. 2011), only one study is available for each of these habitats, whilst no primary studies exist for machair. Dunes (e.g. Beaumont et al. 2010; Landry and Hindsley 2011), coastal lagoons (e.g. Alberini et al. 2007; O'Garra 2012), mudflats (e.g. Andrews et al. 2006; Shepherd et al. 2007), rocky bottoms (Stål et al. 2008; Kenter et al. 2013, 2014), and kelp forests (e.g. Smith and Wilen 2003; Turpie et al. 2003) have also received very little attention in the valuation literature. Reasonably well studied in the international literature are mangroves (e.g. Barbier et al. 2002; Das and Vincent 2009), intertidal wetlands (e.g. Samonte-Tan et al. 2007; Barbier et al. 2013), estuaries (e.g. Milon and Scrogin 2006; Zheng et al. 2009) and seagrasses (Unsworth et al. 2010; Tuya et al. 2014), for each of which at least 10 studies are available. The ecosystem goods and benefits provided by beaches (assessed in 26 % of the studies, e.g. Hynes et al. 2013; Windle and Rolfe 2013), tropical coral reefs (20 %, e.g. Farr et al. 2014; Pascoe et al. 2014) and the coastal shelf (26 %, e.g. Brouwer 2012; Doherty et al. 2014) have been most frequently valued in the academic literature.

Similar to other studies (e.g. Hynes et al. 2013), it is impossible to undertake a meta-regression analysis of all studies and ecosystem services together, because of the limited availability and distribution of value estimates across ecosystem goods and benefits and habitats of Northern European coastal zones. Such meta-analyses can be useful for benefit transfer purposes, with the necessary caution, because value estimates depend on the nature of the study, i.e. the policy context, the valuation method, the sample and the survey design.

**Table 6.1** Overview of number of global studies for each combination of habitat and ecosystem service

| | Products | Sea defence | Erosion prevention | Healthy climate | Tourism and nature watching | Education and research | Aesthetic: property | Spiritual/aesthetic: wild species, seascapes |
|---|---|---|---|---|---|---|---|---|
| **Dunes** | 0 | 4 | 0 | 1 | 3 | 0 | 1 | 0 |
| **Beaches, shingles** | 0 | 3 | 1 | 0 | 47 | 0 | 11 | 9 |
| **Sea cliffs, small islands** | 0 | 0 | 0 | 0 | 1 | 0 | 0 | 0 |
| **Machair** | 0 | 0 | 0 | 0 | 0 | 0 | 0 | 0 |
| **Lagoons** | 2 | 0 | 0 | 0 | 2 | 0 | 0 | 1 |
| **Salt marshes** | 3 | 5 | 0 | 5 | 5 | 0 | 1 | 4 |
| **Mudflats** | 1 | 2 | 0 | 2 | 1 | 0 | 0 | 1 |
| **Mangroves** | 13 | 7 | 3 | 0 | 5 | 0 | 0 | 4 |
| **Inter. wetland** | 2 | 1 | 0 | 0 | 5 | 0 | 1 | 9 |
| **Seagrass beds** | 8 | 0 | 0 | 1 | 2 | 0 | 0 | 3 |
| **Kelp forest** | 3 | 0 | 0 | 0 | 0 | 0 | 0 | 0 |
| **Estuaries, bays, fjords** | 5 | 0 | 0 | 1 | 13 | 0 | 2 | 6 |
| **Tropical coral reefs** | 5 | 0 | 1 | 0 | 44 | 2 | 1 | 17 |
| **Cold water coral reefs** | 0 | 0 | 0 | 0 | 0 | 0 | 0 | 1 |
| **Rocky bottom** | 1 | 0 | 0 | 0 | 1 | 0 | 0 | 0 |
| **Coastal shelf** | 16 | 0 | 0 | 4 | 32 | 0 | 0 | 29 |
| **Open ocean** | 0 | 0 | 0 | 1 | 0 | 0 | 0 | 1 |

Note: The numbers refer to the number of studies that provide at least one value for the ecosystem service in a particular habitat type

### *6.3.2 GB-Based Studies*

The 30 primary GB valuation studies cover various habitats and goods and benefits. Recreational values are most frequently provided in the literature. Table 6.2 provides an overview of the available value estimates.[1] Unless stated otherwise, value estimates in this section are expressed in £, 2012 prices. Original values reported in the original studies have been corrected for inflation, using the National Accounts figures from ONS (last updated 27 March 2013). Two studies were excluded because of limited reliability and validity of the valuation methods (Mangi et al. 2011; Voke et al. 2013), whilst the study by Bateman et al. (2009) did not clearly present value estimate that could be used for benefit transfer purposes. The MPA study by Kenter et al. (2013, 2014) provides a number of generic habitat estimates, but for goods-habitat specific combinations further calculations using the model results are necessary, and therefore not presented here.

**Products** The first category includes goods and benefits of provisioning services. Coastal and marine ecosystems provide not only fish and shellfish for human consumption, fish feed and bait, fertiliser and biofuels, ornaments and aquaria, medicines and biotechnology, but coastal margins are also used for grazing, the collection of wild mushrooms and berries, other crops, reed, timber and seaweed (Jones et al. 2011).[2] Five studies provide primary data for GB. Luisetti et al. (2011) estimate the contribution of created salt marshes in the Blackwater estuary (through coastal realignment schemes) that act as a nursery for species relevant to commercial fisheries using estimates of juvenile bass abundance, average survival rates of fish up to commercial sizes and local market prices. However, fish production functions are highly site-specific and transferring the function from the Blackwater site to another salt marsh would not be reliable (Luisetti et al. 2014). Three studies look at coastal shelf areas (Crilly and Esteban 2013; Austen et al. 2010; Beaumont et al. 2010). These annual gross values cannot be split into values per unit area without data on vessels activities across the coastal waters. Moreover, current harvesting levels may not be sustainable so the current value estimates are of limited use for future projections and scenarios (Beaumont et al. 2010). Fisheries also have other negative externalities, which are not reflected in market prices (Crilly and Esteban 2013).

One study, the CE by Jobstvogt et al. (2014), assesses the WTP for protecting deep sea areas for their option values related to new medicinal products. Respondents were willing to contribute to the creating of deep sea MPAs in Scotland and protect animals with potential for new products if that potential was high.

**'Healthy' Climate** Typically, valuation studies use existing estimates of carbon sequestration rates of coastal and marine ecosystems and apply these to their case study area, combined with existing carbon value estimates. By using different sedimentation rates (Andrews et al. 2000; Adams et al. 2012) and carbon prices

[1] Tropical coral reefs and mangroves are of little importance to GB and therefore not included in Table 6.2.

[2] Recreational extraction of food and other products are included in the tourism category.

**Table 6.2** Overview of UK valuation studies published since 2000 in the academic, peer-reviewed literature

| Ecosystem service | Habitat | Case study and reference | Valuation method | Value in 2012 prices (£/year unless stated otherwise) |
|---|---|---|---|---|
| Products: fisheries (nursery) | Salt marshes | Blackwater: Luisetti et al. (2011) | Market prices | 8.27–12.86/ha (after 5 years) |
| Products: fisheries | GB coast/open sea | UK cod fisheries in North Sea: Crilly and Esteban (2013) | Gross value | 4.4 million |
| | | Fisheries: Beaumont et al. (2010) | Gross value | Fisheries: 619 million |
| | | Fish and shellfish farming: Austen et al. (2010) | Gross value | Fish farms: 364 million |
| | | | | Shellfish farms: 26 million |
| | | Option value of medicinal products: Jobstvogt et al. (2014) | CE | 37.85 |
| Healthy climate | Dunes | GB: Beaumont et al. (2010) | Abatement costs (2010 DECC) | 33-251/ha |
| | Salt marsh and mudflats | Humber: Andrews et al. (2006) | SCC | 14/ha |
| | | Blackwater: Shepherd et al. (2007) | SCC | 13–53/ha |
| | Salt marshes | Blackwater: Luisetti et al. (2011) | Market prices, SCC | 1–865/ha |
| | | GB: Beaumont et al. (2010) | Abatement costs (2010 DECC) | 63–646/ha |
| | Sea grasses | GB: Luisetti et al. (2013) | Abatement costs (2012 DECC), SCC | 103/ha (CI: 6.36–445/ha) |
| | Coastal shelf | GB: Beaumont et al. (2010) | Abatement costs (DECC) | 7 billion (+/−50 %) |
| Coastal erosion prevention | Shingle bank (beach) | Protection of recreational values of Cley Marshes: Bateman et al. (2001) | TC<br>CV | TC: 66/hh/visit<br>CV: 2–81/hh |

(continued)

**Table 6.2** (continued)

| Ecosystem service | Habitat | Case study and reference | Valuation method | Value in 2012 prices (£/year unless stated otherwise) |
|---|---|---|---|---|
| Sea defence | Dunes | England and Wales: Beaumont et al. (2010, 2014) | Replacement costs | England: 181–540 million;<br>Wales: 56 billion |
| | | Sefton Coast: Van der Meulen et al. (2004) | Management costs | 309–1,949/ha |
| | Shingle beaches | England: Beaumont et al. (2010) | Replacement costs | 0.82 billion |
| | Salt marshes and mudflats | Humber: Andrews et al. (2006) | Replacement costs, avoided costs, opportunity costs | Capital costs: 1,033,420/km<br>Opportunity costs: 2,685–3,031/ha<br>Replacement costs savings: 786,623/km<br>Maintenance costs savings: 3,730–4,189/km |
| | | Blackwater: Shepherd et al. (2007) | Avoided costs | Maintenance costs savings: 4,950/km |
| | Salt marshes | GB: Beaumont et al. (2010, 2014) | Replacement costs | Total savings: 5.5–9.7 billion<br>2,225–5,191/m wall for 6 m wide marsh;<br>3,856–6,822/m wall for 80 m wide marsh. |
| | | England: Beaumont et al. (2010) | Net replacement costs | 2.25 billion |
| Tourism and nature watching | All | Coastal recreation: Palmieri et al. this book | Meta-analysis | 4/trip<br>England: 39 million |
| | Beaches | Norfolk EC Bathing Water Directive: Georgiou et al. (2000) | CV | 49/hh |
| | | Norfolk beach replenishment Bateman et al. (2001) | CV | 34–41/hh (local – holiday)<br>Total: 971,640 |
| | | Coastal water quality in Scotland: Hanley et al. (2003) | TC, Contingent behaviour | 0.63/trip (7.66/pp)<br>Total: 1.65 million |
| | | Beach protection: Christie and Gibbons (2011) | CE | Beach safety: 38/hh;<br>Surfing conditions: 16.5/hh |

| | | | | |
|---|---|---|---|---|
| | Cliffs, small islands | Lundy Island Marine Nature Reserve: Chae et al. (2012) | TC | 420–672/trip |
| | Salt marshes | Blackwater managed realignment: Luisetti et al. (2011) | CE | Access: 4.91/hh<br>Access (ln(ha)): 1.36/hh |
| | Estuary | Conservation and regeneration in the inner Firth of Forth: Kenter et al. (2014)[a] | CE | Water quality: 15.28/pp<br>Hide (South of river): 34.77/pp<br>Woodland (South of river): 33.61/pp<br>Woodland (Stirling): 27.28 |
| | Coastal shelf | Seal conservation in England: Bosetti and Pearce (2003) | CV | 10–12/view |
| | | Whale-tourism in West-Scotland: Parsons et al. (2003) | Gross value | Whale-tourism: 2.3 million<br>Total: 7.9 million |
| | | Sea angling in England: Lawrence (2005) | CE | Per day-trip: 6.72–14.93 |
| | | Lyme Bay, England: Rees et al. (2010) | Gross value | Total: 19.8 million<br>Angling: 14.8 million<br>Diving: 1.1 million<br>Wildlife watching: 3.8 million |
| | | Biodiversity related recreation in Wales: Ruiz Frau et al. (2013) | Gross value (financial revenues) | Diving: 8.4 million<br>Kayaking: 2.7 million<br>Boating: 14.5 million<br>Seabird watching: 3.9 million |
| Spiritual and cultural wellbeing and aesthetic benefits of wild species and seascapes | Salt marshes | Blackwater: Luisetti et al. (2011) | CE | Additional bird species: 2.09–4.06/hh |
| | Intertidal wetlands | Otter and bird protection: Birol and Cox (2007) | CE | Otter hold creation: 37.19/pp<br>Protection birds: 1.41/pp |
| | Estuary | Conservation in the inner Firth of Forth: Kenter et al. (2014)[a] | | Bird population +1,000: 3.25/pp<br>Prevention 1 local bird species: 17.60 pp |

(continued)

**Table 6.2** (continued)

| Ecosystem service | Habitat | Case study and reference | Valuation method | Value in 2012 prices (£/year unless stated otherwise) |
|---|---|---|---|---|
| | Coastal shelf | MPAs in the UK: McVittie and Moran (2010) | CE | Halting loss of biodiversity and ES:<br>England: 75/hh<br>Wales: 116/hh<br>Scotland: 23/hh,<br>Northern Ireland: 37/hh<br>Increasing biodiversity<br>England: 75/hh<br>Wales: 66/hh<br>Scotland: 26/hh<br>Northern Ireland: 41/hh |
| | | Marine species conservation: Ressurreicao et al. (2011, 2012) | CE | All one-off payments<br>Mammals: 43–49/hh<br>Birds: 39–44/hh<br>Fish 38–43/hh<br>Invertebrates: 36–41/hh<br>Algae: 46–53/hh |
| | | Seal conservation in England: Bosetti and Pearce (2003) | CV | Total: 689,239 |
| | | Species protection in Scottish deep sea: Jobstvogt et al. (2014) | CE | Intermediate level (+300 species): 26.28<br>High level (+600 species): 38.70 |

[a]Values included here for Kenter et al. (2013, 2014) are individual non-deliberative values

ranging from £4 to £230/tC, Luisetti et al. (2011) show that the value of the carbon storage capacity by salt marsh re-creation projects may vary from £1 to £865/ha/year. A similarly wide range is presented in Beaumont et al. (2010, 2014): from £63 to £646/ha/year. Two studies (Andrews et al. 2006; Shepherd et al. 2007) look at the carbon sequestration by salt marshes and mudflats in the Blackwater and Humber catchments using the average concentrations of particulate C, N and P from Andrews et al. (2000) and Jickells et al. (2003). Estimates are also available for dunes (Beaumont et al. 2010, 2014) based on a carbon sequestration study by Jones et al. (2008) and carbon prices from UK DECC; and sea grass (*Zostera marina* species) (Luisetti et al. 2013). Beaumont et al. (2014) provide a figure for machair but since this estimate is based on sand dune grasslands and not on primary biophysical research in machair areas, this is excluded from our overview in Table 6.2. The sequestration rates for dunes, sea grasses, salt marshes and mudflats used in the studies are comparable to other studies elsewhere and are considered to be transferable across space and time (Luisetti et al. 2014).

The carbon sequestration of marine habitats through primary production of phytoplankton has been assessed, but its net contribution to the reduction of atmospheric $CO_2$ levels depends on the transportation of carbon to deep oceans where carbon is stored permanently (see Heckbert et al. 2011). In GB (coastal) shelf seas it is unlikely that this carbon will be transported to the deep ocean. Nevertheless, Beaumont et al. (2010) estimate that in 2004, the value of carbon sequestration in marine habitat by phytoplankton based on primary production was £7 billion/year.

**Prevention of Coastal Erosion** Natural habitats play an important role in coastal protection policies in GB. Coastal protection can be provided in terms of the prevention of coastal erosion when the gradual loss of land is mitigated by coastal habitats, or in terms of sea defence that reduce the risk of sea flooding and inundation related to natural hazards (see also Section 11.3.2.1 in Jones et al. 2011). Coastal protection values include benefits of ecosystem services provided by areas that are prevented from being lost through the protection provided by coastal margins.

The value of coastal erosion prevention includes avoided losses of property, agriculture, recreational uses etc. that take place without erosion. Bateman et al. (2001) report on the only GB-based study on benefits of coastal erosion prevention published since 2000. They address the recreational values of the freshwater Cley Marshes Natural Reserve that are protected from saltwater inundation by a shingle bank, using a combined TC-CV survey. The results of the study show that the aggregate annual recreational benefits are around £786,000 – £1,970,000, depending on the welfare estimate used (TC or CV) and the estimated number of visitors to the site, and much higher than the maintenance cost of £30,000–£50,000/year. Limitations of this study include the small sample size, the limited detail on TC and CV WTP functions or analysis, and the imprecise CV scenario description. The usefulness of this study for benefit transfer (BT) may be limited to cases where shingle beaches protect freshwater marshes.

**Sea Defence** Sea defence values relate to a risk reduction of flood, storm or tidal surge events that would damage infrastructure, business, the natural and historic environment, and other property, and also the risk of life. This risk reduction benefit depends on the location, depth and flow rate of the potential flood event. Two existing meta-analyses have not found significantly higher values for storm protection provided by wetlands (Brander et al. 2006) or lagoons (Enjolras and Boisson 2010), but these results do not necessarily imply that these habitats do not provide sea defence services. The benefits of sea defence have been assessed for several ecosystems in GB: marshes, mudflats, mangroves, beaches and dunes. All studies use cost-based valuation methods. The main limitation of these cost-based estimates is that they do not reflect the value of the goods and benefits protected by ecosystem sea defence, including values of commercial and residential properties, agriculture and recreation. They are typically a lower bound estimate of society's willingness-to-pay. Moreover, the costs of managed realignment vary widely across sites (Tinch and Ledoux 2006).

Salt marshes allow for building lower man-made sea walls, or no walls at all. Andrews et al. (2006) and Shepherd et al. (2007) estimate that replacing hard defences by salt marshes and mudflats would provide savings on replacement costs of unsatisfactory hard defence and maintenance costs. Since salt marshes and mudflats also provide societal benefits through carbon sequestration, recreational opportunities and their nursery function, the overall cost-benefit ratio supports the implementation of this soft approach to coastal defence when viewed over >25 year time scales. The resulting cost savings vary depending on the width of the salt marsh beside the sea wall. The GB-wide figures presented in Beaumont et al. (2010), based on cost-data from King and Lester (1995), for replacing salt marshes with man-made sea defences ignore the width of the salt marsh.

The total replacement cost of shingle shores in England are estimated at £0.82 billion, whilst sand dunes defence services are worth £0.54 billion (see Beaumont et al. 2010, also for limitations) and lower when using an alternative approach based on Pye et al. (2007): £181 million in England and £56 million in Wales. However, the latter are very conservative estimates and only apply to dunes without any additional artificial defence structures near high value land. The study by Van der Meulen et al. (2004) addresses the management costs of two dune sites on the Sefton Coast, one which is managed as a Nature Reserve and a busier one managed as a semi-park. However, these costs are not only for sea defence, as these dunes are also managed for their recreational use and cultural, spiritual, and aesthetic (biodiversity, non-use) benefits, but it is not possible to assign separate values to each of these benefit categories.

**Tourism and Nature Watching** There are many international studies on the benefits of tourism for beaches, tropical coral reefs and coastal shelf areas, yet no value estimates for open oceans, machair and cold coral reefs. Palmieri et al. (Chap. 12) use the results of Sen et al. (2014) to estimate the recreational values of coastal areas in the England. Based on an estimated £4 per trip, the total benefits amount to £39 million. However, these values cannot be assigned to specific habitats and the value per trip is based on an international recreation meta-analysis. Habitat-specific studies are available for beaches, small islands, salt marshes and the coastal shelf in

GB. Three studies are available that assess values associated with beach recreation more locally. Georgiou et al. (2000) use an open-ended CV survey to estimate public WTP for achieving compliance with the EC Bathing Water Directive to ensure safe bathing conditions at beaches in East Anglia. Hanley et al. (2003) combine TC and Contingent Behaviour data to estimate the WTP for better coastal water quality at seven different beaches in Scotland. The results suggest that the number of trips would increase should water quality improve to 'very good' standards, with associated aggregate benefits of £1.65 million/year. Bateman et al. (2001) assess the benefits of beach replenishment to avoid coastal erosion – and thereby obtain extra recreational possibilities in Caister-on-Sea, Norfolk, using an open-ended CV survey. The resulting aggregate benefits of £971,640/year would outweigh the cost of beach replenishment. Although these three primary studies fulfil most standard reliability and validity criteria, the surveys were executed prior to 2000 and the use of these values in BT may produce less reliable results. A more recent, but rather specific CE study on beach amenities assessed WTP for a change in coastal defences in Borth, North Wales (Christie and Gibbons 2011). These results could be applied to similar interventions that improve beach safety and surfing conditions.

One GB-based study falls into the small islands category. Chae et al. (2012) use TC to estimate the non-market recreational benefits arising from the Lundy Island Marine Nature Reserve. The estimated mean WTP for visiting Lundy is high compared to other studies. This may be because of the protected and unique status of Lundy, but also because of the inclusion of multipurpose trips or the small sample. The CE presented in Luisetti et al. (2011) of salt marshes shows that respondents attribute higher welfare to salt marshes that are accessible for recreation. WTP estimates decrease with distance and increase with the size of the marsh in a non-linear way. Two studies assess marginal values for recreational activities at the coastal shelf. Bosetti and Pearce (2003) use a CV study to assess the use value of seal conservation in southwest England, but the resulting values are difficult to relate to marginal increases in seal populations. The results of the CE about recreational coastal angling in southwest England presented in Lawrence (2005) show that WTP values per fishing trip varied by species. The relationship between catch size and WTP is non-linear (declining), and increasing the size of individual fish would have a larger impact on WTP than increasing the catch per day in this study. These results can be used in scenarios of change, as they reflect the values associated with specific changes in biophysical parameters.

Tourism values are were assessed within the UK NEA FO on the benefits of Marine Protected Areas by Kenter et al. (2013, 2014) for a range of substrate/ habitats, including rocky seafloors with shell beds, large kelp, seaweeds and sea-pens, and sandy and muddy sea floors with different types of plant growth, including soft corals and sponges, as well as estuarine areas. These habitats cover a range of features of conservation interest (FOCI, see Chap. 8). The combined CE-CV study provided positive WTP estimates for MPA development, which vary across habitats. Positive WTP values are also found for sites where seals, octopus and birds may be encountered. WTP was also higher in both the CV and CE exercise for sites that were accessible by shore, boat and pier, whilst access out at sea or where boat use is prohibited were associated with negative effects. Size did not have any impact on recreational values, but distance was significant and negative. A limitation of the study is the use of a voluntary donation as payment vehicle, which is generally considered not to be incentive compatible. The study compared individual and deliberative approaches to valuation, and found that values would generally decrease after deliberation, which may be because the deliberative results were based on a 'fair price' whilst the individual WTP questions aim to elicit maximum individual WTP.

Kenter et al. (2014) also present the results of a CE study in the Firth of Forth, an estuary in Scotland. The results show significant positive WTP for improvement of water quality, an increase of the bird populations, the presence of a hide (but only South of Forth) and new woodland planted (but only South of Forth and near Stirling – there are many woodland North of the river). Again, this study found that deliberation resulted in lower WTP values, both at individual level as well as when WTP was expressed by the group as a fair price.

Three studies assess the direct income earned in the coastal shelf from tourism and recreation (Parsons et al. 2003; Rees et al. 2010, Ruiz Frau et al. 2013). Although these values indicate the economic importance of coastal recreation, the estimates are not directly related to changes in environmental quality or habitat extent and their use in scenario analysis would require additional assumptions. Moreover, they do not reflect consumer surplus, i.e. the welfare that people derive from coastal and marine tourism on top of what they have to pay on accommodation, transport, excursions, entrance fees, etc.

**Aesthetic Values as Reflected in Property Prices** Cultural values range from use values related to tourism and nature watching, aesthetic values, education and research, to goods and benefits of spiritual and cultural wellbeing. Aesthetic benefits are sometimes reflected in property values when people are willing to pay an additional price in the housing market that can be attributed to the presence of nearby environmental amenities. The only GB-based study has been developed for the UK NEA 2011. Mourato et al. (2010) find that house prices in England are not significantly associated with distance to the coastline or the availability of marine and coastal margins in the $km^2$ in which a house is located. However, it may be that the effect of seascape aesthetics on housing prices could not picked up at the coarse scale of this analysis and should not be considered conclusive evidence for the absence of aesthetic benefits reflected in GB housing prices. International studies (n = 17), mostly from the USA, find evidence of the added value of nearby ecosystem services in house prices. Given the large differences in housing markets between countries, transferring values to GB is expected to generate large errors in value estimates (see Sect. 6.4).

**Spiritual and Cultural Well-Being and Aesthetic Benefits of Wild Species and Seascapes** There are over 60 international valuation studies that address the economic welfare that people derive from biodiversity, species, habitat and/or landscape conservation. These reflect both spiritual and cultural wellbeing and aesthetic values. Seven SP studies provide primary value estimates for GB.

The study by Luisetti et al. (2011) on the WTP for salt marsh creation along the English coast also assessed WTP per observable protected bird species which, at least in part, non-use values. Marginal WTP is declining as the number of species increases, from £2.09/hh/year for three additional species to £4.06/hh/year for five additional species. The study also shows that people are willing to pay for salt marsh creation even when they won't be allowed access to the site. Birol and Cox (2007) use a CE to assess the WTP for otter hold creation and protected bird species in wetlands. The sample contained both users and non-users, and was small, and the models relatively simple. Hence, reliable extraction of pure non-use values from these studies is not possible. The results of the CE in the estuary Firth of Forth by Kenter et al. (2014) reveal a positive WTP for preventing a local species from extinction, in addition to the value of increasing the bird population in general.

In their CV study on seal conservation in southwest England, Bosetti and Pearce (2003) found respondents willing to pay to mitigate conflicts between fishermen and seals and conserve seals in the wild. However, besides the relatively small sample, the payment vehicle employed for non-use values in this study (voluntary donation) is not considered to be incentive compatible, because they could avoid actual payments would the proposed donation request be implemented.

McVittie and Moran (2010) use a CE to ask respondents for their WTP to install Marine Protected Areas (MPAs) in the coastal waters of England, Wales, Northern Ireland and Scotland. Part of the WTP values reflect use values. The levels of the attributes were defined as 'increase biodiversity' and 'halt loss of biodiversity', hence the change in ecosystem service provision is not described quantitatively (mainly because a lack of such information), which may limit the possibilities for BT. Ressurreicao et al. (2011, 2012) implemented a CV survey to assess the WTP for marine species among residents and visitors in three European coastal areas, including the Isles of Scilly. The results show that the absolute WTP for the prevention of species loss are around 2–3 % of monthly household income. The results did not show significant sensitivity to scope, i.e. losing fewer species was not associated with significantly higher WTP, which may be due to warm glow effects or limited understanding about the implications of species loss and ecological uncertainty about the effects of species loss on other communities. Kenter et al. (2013, 2014) also find a positive WTP for protection of various marine landscapes under MPA regulations, symbolic sealife species, and for the protection of vulnerable marine species that anglers or divers would normally not encounter.

The CE by Jobstvogt et al. (2014) to assess the WTP for the conservation of deep sea organisms in Scotland shows that, despite limited knowledge about deep sea biodiversity, respondents were willing to contribute to MPAs in deep sea areas.

**Education and Research** No academic papers present values of education and research. Financial values are available from UKMMAS (2010) and Pugh and Skinner (2002), as reported in Beaumont et al. (2008). There are only two other, non-GB, academic studies that meet our study selection criteria and assess the economic value of education and research (Cesar and van Beukering 2004; Samonte Tan et al. 2007). While these studies are not directly applicable to GB they do provide some notion of the magnitude of this category of benefit.

## 6.4 Prioritisation of Future Research Resources

The gaps in the primary GB-based valuation literature limit the possibilities to inform management, especially for ecosystem goods and benefits and habitats that are considered to be important. There are no GB valuation studies for a number of the habitats (machair, coastal lagoons, cold water corals and open oceans) published in the academic literature since 2000. There are also no value estimates for amenity effects on property values and education and research, only one study on the benefits of prevention of coastal erosion, and only values for product provisioning in salt marshes and the coastal zone. Moreover, the available studies use different valuation methods, and the results are not necessarily comparable and vary in terms of their reliability and validity.

We compare the availability of existing valuation studies to expert-based judgements on the importance of coastal and marine habitats and the ecosystem goods and benefits they provide. The UK NEA 2011 provides an assessment of the importance of the different types of coastal margins in terms of their contribution to human wellbeing of the various goods and benefits (or the amount of good/benefit delivery per unit area) that these habitats provide (Jones et al. 2011). We complement this with a comparable importance matrix for marine habitats in GB, developed in an expert-workshop during the UK NEA Follow-On project.

Table 6.3 presents the results: the number in each cell reflects the number of studies that are available for that particular good/benefit in the habitat, and the colour coding reflects the availability-importance score.

As the many red and orange cells in Table 6.3 indicate, there are considerable gaps in the GB valuation literature related to ecosystem goods and benefits provided by coastal ecosystems deemed important by experts. Cultural values (here under education, research, spiritual and aesthetic values of wild species and seascapes) are poorly represented in the monetary valuation studies literature despite the service-habitat combination being deemed important, and this holds to a lesser extent for cultural use values related to recreation. Sea defence and carbon sequestration benefits of coastal habitats have received little recent attention despite the significant risks that climate change and sea level rise may pose. No carbon sequestration valuation studies are available for sea cliffs and small islands, machair, lagoons, intertidal wetlands, estuaries, kelp forests and cold water coral reefs, whilst studies for the coastal shelf and the open ocean are associated with large uncertainties about the longer-term storage.

Provisioning services related to land-based activities on coastal margins, including the production of crops, meat, wild food, wool, reed, grasses, timber and turf, require more attention. No GB studies on products are available other than those on (shell-) fisheries and aquaculture and no studies exist for dunes, machair, mudflats, seagrass beds, kelp forest, estuaries, cold water corals, rocky bottoms and the open ocean. The studies in our global valuation dataset also do not provide value estimates for these goods and benefits from coastal habitats in countries like the UK. Future studies should also provide more insight into sustainable harvesting

**Table 6.3** Importance of ecosystem services per coastal habitat and the availability of UK-based valuation studies

| | Products | Sea defence | Erosion prevention | Healthy climate | Tourism and nature watching | Education research | Aesthetic: property[a] | Spiritual/ aesthetic: wild species, seascapes |
|---|---|---|---|---|---|---|---|---|
| Dunes | 0 | 2 | 0 | 1 | 1 | 0 | 0 | 0 |
| Beaches | 0 | 1 | 1 | 0 | 3 | 0 | 0 | 0 |
| Sea cliffs | 0 | 0 | 0 | 0 | 1 | 0 | 0 | 0 |
| Machair | 0 | 0 | 0 | 0 | 0 | 0 | 0 | 0 |
| Lagoons | 0 | 0 | 0 | 0 | 0 | 0 | 0 | 0 |
| Salt marshes | 1 | 3 | 0 | 4 | 1 | 0 | 0 | 1 |
| Mudflats | 0 | 2 | 0 | 2 | 1 | 0 | 0 | 0 |
| Inter. wetland | 0 | 0 | 0 | 0 | 0 | 0 | 0 | 1 |
| Seagrass beds | 0 | 0 | 0 | 1 | 1 | 0 | 0 | 1 |
| Kelp forest | 0 | 0 | 0 | 0 | 1 | 0 | 0 | 1 |
| Estuaries | 0 | 0 | 0 | 0 | 2 | 0 | 0 | 2 |
| Cold water coral reefs | 0 | 0 | 0 | 0 | 0 | 0 | 0 | 0 |
| Rocky bottom | 0 | 0 | 0 | 0 | 1 | 0 | 0 | 1 |
| Coastal shelf | 3 | 0 | 0 | 1 | 6 | 0 | 0 | 5 |
| Open ocean | 0 | 0 | 0 | 0 | 0 | 0 | 0 | 0 |

Red: services of high importance with no relevant UK valuation studies
Orange: services of high importance with one UK valuation study, or services of medium importance with no UK valuation studies
Yellow: services of high importance with two or more UK valuation studies, or services of medium importance with one UK valuation study
White: services of low importance or services of medium importance with two or more UK valuation studies
[a]Property related aesthetic values are not included in Table 11.3 of UK NEA 2011

levels, analyse the value of fisheries net of other capital inputs, and include the economic value of other raw materials, including seaweed and pharmaceuticals.

It is also remarkable that are no studies for the (flow of) goods and services provided by machair, even though this is a unique type of habitat and only found in the UK and Ireland, and considered to be very important for sea defence, recreation, education, cultural wellbeing, aesthetics and biodiversity. Cold water coral reefs have not been addressed in the UK yet; the study by Jobstvogt et al. (2014) assesses the option and biodiversity values of deep seas in the Scottish EEZ.

More valuation efforts should be directed towards intertidal wetlands and estuaries. Their provision of products and different cultural services (tourism, education and research, aesthetic values of species and seascapes) are considered to be important in terms of their contribution to human wellbeing. For estuaries, intertidal wetlands and other 'habitat complexes' or 'habitat mosaics', it may be possible to use valuation studies for the habitat types that are present in the estuary (or habitat mosaic) of interest. However, the biophysical ecosystem service provision level as well as the economic values for the associated benefits may not be independent from the adjacent habitats within a habitat mosaic. In the presence of synergistic or antagonistic effects of one habitat type, fragmented within the mosaic, on the delivery of any particular service from another interspersed habitat type may not have the same value as a single block of habitat of equivalent overall size.

Benefit transfer approaches could help to fill some of the gaps. Table 6.1 shows that for some of the goods and benefits for which there exist no primary GB studies, value estimates from other countries may be available. As a rule of thumb, we suggest that for benefit transfer to the UK using international studies, studies from North- and West-Europe could be applied with the necessary caution, then studies from South- and East-Europe with more caution, followed by Australian and North-American studies with further increased caution, and studies from elsewhere should probably not be applied due to large differences in cultural, economic and ecological differences. There are four North- and West-Europe studies published that provide values for habitat/good and benefit combination for which no UK studies are available, which we will mention here but not evaluate. Nunes and Van den Bergh (2004) present a TC-CV study on the WTP to protect beaches in the Netherlands against algae blooms. Meyerhoff (2004) presents a CV study in Germany on the tourism benefits of the Wadden Sea. Stål et al. (2008) present a study on fisheries and the nursery function supporting commercial fisheries provided by seagrass beds and rocky bottom areas in Sweden. These studies may provide an initial figure of the order of magnitude of values of the goods and benefits but are likely to arise in high errors given the differences in social and ecological characteristics and are probably insufficiently reliable for socially efficient and equitable decision making.

It is difficult to prioritise research efforts based on national or international policy needs based on habitats or ecosystem goods and benefits, such as the OSPAR convention, the WFD and MSFD, and Strategic Environmental Assessments and Environmental Assessment regulations (see Chap. 1). The WFD and MSFD together cover all coastal and marine habitats and therefore economic value estimates are required for all types of habitats for impact assessments of measures. Similarly, the UK Biodiversity Action Plan (UKBAP) has defined 24 priority habitats and valuation information may be useful for all of these.

## 6.5 Concluding Remarks

Clear gaps have been identified in this review exercise for both the international and the UK coastal and marine ecosystem valuation data. A number of important habitats, ecosystem services and related goods and benefits have few or no valuation estimates assigned to them. While benefits transfer may offer some pragmatic assistance to cover a limited number of the gaps, this procedure is unlikely to be any sort of panacea. Both temporal and cultural bias constraints remain formidable challenges for any benefits transfer exercise using data more than a decade old and spatially more distant than a rough boundary around Northern Europe. The only real exceptions to this rule are global benefits such as those related to carbon sequestration and storage.

The obvious conclusion form this review analysis is that more primary valuation research needs to be undertaken. Table 6.3 offer some guidance on the foci for this possible new research programme for the UK. Highlighted gaps include the sea defence and coastal erosion prevention benefits, as well as climate benefits and provisioning services (products) provided by coastal habitats. For marine ecosystem services, more valuation studies may be required for aesthetic values and spiritual and cultural wellbeing from seascapes and wild species diversity, as well as products and other raw materials, education and research. Finally, the complexity of 'mosaic' habitats, such as intertidal wetlands and estuaries, may require valuation studies that consider these in aggregate terms, rather than trying to disentangle the values goods and benefits provided by sub-habitat types independently and at the same time avoiding double counting.

## References

Adams, C. A., Andrews, J. E., & Jickells, T. D. (2012). Nitrous oxide and methane fluxes vs carbon and nutrient burial in new intertidal and salt marsh sediments. *Science of the Total Environment, 434*, 240–251.

Alberini, A., Zanatta, V., & Rosato, P. (2007). Combining actual and contingent behavior to estimate the value of sports fishing in the Lagoon of Venice. *Ecological Economics, 61*(2), 530–541.

Andrews, J. E., Samways, G., Dennis, P. F., & Maher, B. A. (2000). Origin, abundance and storage of organic carbon and sulphur in the Holocene Humber Estuary – Emphasising human impact on storage changes. In I. Shennan & J. E. Andrews (Eds.), *Holocene land-ocean interaction and environmental change around the North Sea* (Geological Society Special Publication No. 166, pp. 145–170).

Andrews, J. E., Burgess, D., Cave, R. R., Coombes, E. G., Jickells, T. D., Parkes, D. J., & Turner, R. K. (2006). Biogeochemical value of managed realignment, Humber estuary, UK. *Science of the Total Environment, 371*(1), 19–30.

Austen, M., Malcolm, S., Frost, M., Hattam, C., Mangi, S., Mieszkowska, N., Stentford, G., Burrows, M., Butenschön M., Merino, G., Miles, A., & Smyth T. (2010). *Chapter 10: Marine habitats*. UK National Ecosystem Assessment. Cambridge: UNEP-WCMC.

Barbier, E. B., Strand, I., & Sathirathai, S. (2002). Do open access conditions affect the valuation of an externality? Estimating the welfare effects of mangrove-fishery linkages in Thailand. *Environmental and Resource Economics, 21*(4), 343–365.

Barbier, E. B., Georgiou, I. Y., Enchelmeyer, B., & Reed, D. J. (2013). The value of wetlands in protecting southeast Louisiana from hurricane storm surges. *PloS One, 8*(3), e58715.

Bateman, I. J., Turner, R. K., Klein, R. J. T., & Langford, I. H. (2001). The application of the cost benefit method to sea defence and coastal protection management in England. In R. K. Turner, I. J. Bateman, & W. N. Adger (Eds.), *Economics of coastal and water resources: Valuing environmental functions* (pp. 113–142). Dordrecht: Kluwer.

Bateman, I. J., Day, B. H., Jones, A. P., & Jude, S. (2009). Reducing gain-loss asymmetry: A virtual reality choice experiment valuing land use change. *Journal of Environmental Economics and Management, 58*, 106–118.

Beaumont, N. J., Austen, M. C., Mangi, S. C., & Townsend, M. (2008). Economic valuation for the conservation of marine biodiversity. *Marine Pollution Bulletin, 56*, 386–396.

Beaumont, N., Hattam, C., Mangi, S., Moran, D., van Soest, D., Jones, L., & Tobermann, M. (2010). *Economic analysis of ecosystem services provided by UK coastal margin and marine habitats* (Final report to the economics team of the UK National Ecosystem Assessment). Plymouth: Plymouth Marine Laboratory.

Beaumont, N. J., Jones, L., Garbutt, A., Hansom, J. D., & Toberman, M. (2014). The value of carbon sequestration and storage in coastal habitats. *Estuarine, Coastal and Shelf Science, 137*, 32–40.

Birol, E., & Cox, V. (2007). Using choice experiments to design wetland management programmes: The case of Severn Estuary wetland, UK. *Journal of Environmental Planning and Management, 50*(3), 363–380.

Bosetti, V., & Pearce, D. (2003). A study of environmental conflict: The economic value of Grey Seals in southwest England. *Biodiversity and Conservation, 12*, 2361–2392.

Brander, L. M., Florax, R. J. G. M., & Vermaat, J. E. (2006). The empirics of wetland valuation: A comprehensive summary and a meta-analysis of the literature. *Environmental and Resource Economics, 33*, 223–250.

Brander, L. M., Van Beukering, P., & Cesar, H. S. (2007). The recreational value of coral reefs: A meta-analysis. *Ecological Economics, 63*(1), 209–218.

Brander, L., Wagtendonk, A. J., Hussain, S. S., McVittie, A., Verburg, P. H., de Groot, R. S., & van der Ploeg, S. (2012). Ecosystem service values for mangroves in Southeast Asia: A meta-analysis and value transfer application. *Ecosystem Services, 1*(1), 62–69.

Brouwer, R. (2000). Environmental value transfer: State of the art and future prospects. *Ecological Economics, 32*, 137–152.

Brouwer, R. (2012). Constructed preference stability: A test–retest. *Journal of Environmental Economics and Policy, 1*(1), 70–84.

Cesar, H. S., & Beukering, P. V. (2004). Economic valuation of the coral reefs of Hawai'i. *Pacific Science, 58*(2), 231–242.

Chae, D. R., Wattage, P., & Pascoe, S. (2012). Recreational benefits from a marine protected area: A travel cost analysis of Lundy. *Tourism Management, 33*, 971–977.

Christie, M., & Gibbons, J. (2011). The effect of individual 'ability to choose' (scale heterogeneity) on the valuation of environmental goods. *Ecological Economics, 70*, 2250–2257.

Crilly, R., & Esteban, A. (2013). Small versus large-scale, multi-fleet fisheries: The case for economic, social and environmental access criteria in European fisheries. *Marine Policy, 37*, 20–27.

Das, S., & Vincent, J. R. (2009). Mangroves protected villages and reduced death toll during Indian super cyclone. *Proceedings of the National Academy of Sciences, 106*(18), 7357–7360.

Doherty, E., Murphy, G., Hynes, S., & Buckley, C. (2014). Valuing ecosystem services across water bodies: Results from a discrete choice experiment. *Ecosystem Services, 7*, 89–97.

Enjolras, G., & Boisson, J. M. (2010). Valuing lagoons using a metaanalytical approach: Methodological and practical issues. *Journal of Environmental Planning and Management, 53*(8), 1031–1049.

Farr, M., Stoeckl, N., & Alam Beg, R. (2014). The non-consumptive (tourism) 'value' of marine species in the Northern section of the Great Barrier Reef. *Marine Policy, 43*, 89–103.

Georgiou, S., Bateman, I. J., Langford, I. H., & Day, R. J. (2000). Coastal bathing water health risks: Assessing the adequacy of proposals to amend the 1976 EC directive. *Risk, Decision and Policy, 5*(1), 49–68.

Ghermandi, A., & Nunes, P. A. (2013). A global map of coastal recreation values: Results from a spatially explicit meta-analysis. *Ecological Economics, 86*, 1–15.

Hanley, N., Bell, D., & Alvarez-Farizo, B. (2003). Valuing the benefits of coastal water quality improvements using contingent and real behaviour. *Environmental and Resource Economics, 24*, 273–285.

Heckbert, S., Costanza, R., Poloczanska, E. S., & Richardson, A. J. (2011). 12.10 – Climate regulation as a service from estuarine and coastal ecosystems. In E. Wolanski & D. McLusky (Eds.), *Treatise on estuarine and coastal science* (pp. 199–216). Waltham: Academic.

Hynes, S., Tinch, D., & Hanley, N. (2013). Valuing improvements to coastal waters using choice experiments: An application to revisions of the EU Bathing Waters Directive. *Marine Policy, 40*, 137–144.

Jickells, T. D., Andrews, J. E., Cave, R. R., & Parkes, D. J. (2003). Biogeochemical value of intertidal areas – A case study of the Humber estuary. In L. Ledoux (Ed.), *Wetland valuation: State of the art and opportunities for further development* (pp. 61). Norwich, UK: CSERGE.

Jobstvogt, N., Hanley, N., Hynes, S., Kenter, J., & Witte, U. (2014). Twenty thousand sterling under the sea: Estimating the value of protecting deep-sea biodiversity. *Ecological Economics, 97*, 10–19.

Jones, M. L., Sowerby, A., Williams, D. L., & Jones, R. E. (2008). Factors controlling soil development in sand dunes: Evidence from a coastal dune soil chronosequence. *Plant and Soil, 307*(1), 219–234.

Jones, L., Angus, S., Cooper, A., Doody, P., Everard, M., Garbutt, A., Gilchrist, P., Hansom, J., Nicholls, R., Pye, K., Ravenscroft, N., Rees, S., Rhind, P., & Whitehouse, A. (2011). *Chapter 11: Coastal margin habitats*. National Ecosystem Assessment. Cambridge: UNEP-WCMC.

Kenter, J. O., Bryce, R., Davies, A., Jobstvogt, N., Watson, V., Ranger, S., Solandt, J. L., Duncan, C., Christie, M., Crump, H., Irvine, K. N., Pinard, M., & Reed, M. S. (2013). *The value of potential marine protected areas in the UK to divers and sea anglers*. Cambridge: UNEP-WCMC.

Kenter, J.O., Reed, M.S., Irvine, K.N., O'Brien, L., Brady, E., Bryce, R., Christie, M., Church, A., Cooper, N., Davies, A., Evely, A., Everard, M., Fazey, I., Hockley, N., Jobstvogt, N., Molloy, C., Orchard-Webb, J., Ravenscroft, N., Ryan, M., & Watson, V. (2014). *Shared, plural and cultural values of ecosystems*. Work package 5 Report UKNEA. Cambridge, UK: UNEP-WCMC.

King, S. E., & Lester, J. N. (1995). The value of saltmarsh as a sea defence. *Marine Pollution Bulletin, 30*, 180–189.

Landry, C. E., & Hindsley, P. (2011). Valuing beach quality with hedonic property models. *Land Economics, 87*(1), 92–108.

Latinopoulos, D. (2010). Valuing the services of coastal ecosystems: A meta-analysis of contingent valuation studies. *International Journal of Sustainable Development and Planning, 5*, 17.

Lawrence, K. S. (2005). Assessing the value of recreational sea angling in South West England. *Fisheries Management and Ecology, 12*, 369–375.

Londoño, L. M., & Johnston, R. J. (2012). Enhancing the reliability of benefit transfer over heterogeneous sites: A meta-analysis of international coral reef values. *Ecological Economics, 78*, 80–89.

Luisetti, T., Turner, R. K., Bateman, I. J., Morse-Jones, S., Adams, C., & Fonseca, L. (2011). Coastal and marine ecosystem services valuation for policy and management: Managed realignment case studies in England. *Ocean & Coastal Management, 54*, 212–224.

Luisetti, T., Jackson, E., & Turner, R. K. (2013). Valuing the European 'coastal blue carbon' storage benefit. *Marine Pollution Bulletin, 72*(1–2), 101–106.

Luisetti, T., Jickells, T., Andrews, J., Elliot, M., Malcolm, S., Schaafsma, M., Turner, R. K., Watts, W., Burdon, D., Adams, C., & Beaumont, N. (2014). Coastal zone ecosystem services: From

science to values and decision making – case studies. *Science of the Total Environment, 493*, 682–693.
Mangi, S. C., Davis, C. E., Payne, L. A., Austen, M. C., Simmonds, D., Beaumont, N. J., & Smyth, T. (2011). Valuing the regulatory services provided by marine ecosystems. *Environmetrics, 22*, 686–698.
Martín-López, B., Montes, C., & Benayas, J. (2008). Economic valuation of biodiversity conservation: The meaning of numbers. *Conservation Biology, 22*(3), 624–635.
McVittie, A., & Moran, D. (2010). Valuing the non-use benefits of marine conservation zones: An application to the UK Marine Bill. *Ecological Economics, 70*, 413–424.
Meyerhoff, J. (2004). Non-use values and attitudes: Wetlands threatened by climate change. In M. Getzner, C. L. Spash, & S. Stagl (Eds.), *Developing alternatives for valuing nature*. London: Routledge.
Milon, J. W., & Scrogin, D. (2006). Latent preferences and valuation of wetland ecosystem restoration. *Ecological Economics, 56*(2), 162–175.
Moss, D. (2008). EUNIS habitat classification – A guide for users. European topic centre on biological diversity. http://eunis.eea.europe/eu
Mourato, S., Atkinson, G., Collins, M., Gibbons, S., MacKerron, G., & Resende, G. (2010). *Economic assessment of ecosystem related UK cultural services* (The economics team of the UK National Ecosystem Assessment). London: London School of Economics.
Murillas-Maza, A., Virto, J., Gallastegui, M. C., González, P., & Fernández-Macho, J. (2011). The value of open ocean ecosystems: A case study for the Spanish exclusive economic zone. *Natural Resources Forum, 35*(2), 122–133.
Nunes, P. A., & van den Bergh, J. C. (2004). Can people value protection against invasive marine species? Evidence from a joint TC–CV survey in the Netherlands. *Environmental and Resource Economics, 28*(4), 517–532.
O'Garra, T. (2012). Economic valuation of a traditional fishing ground on the coral coast in Fiji. *Ocean & Coastal Management, 56*, 44–55.
Parsons, E. C. M., Warburton, C. A., Woods-Ballard, A., Hughes, A., & Johnston, P. (2003). The value of conserving whales: The impacts of cetacean-related tourism on the economy of rural West Scotland. *Aquatic Conservation of Marine and Freshwater Ecosystems, 13*(5), 397–415.
Pascoe, S., Doshi, A., Thébaud, O., Thomas, C. R., Schuttenberg, H. Z., Heron, S. F., Setiasih, N., Tan, J. C. J., True, J., Wallmo, K., Loper, C., & Calgaro, E. (2014). Estimating the potential impact of entry fees for marine parks on dive tourism in South East Asia. *Marine Policy, 47*, 147–152.
Posford Duvivier Environment. (1996). Financial values of five important marine/coastal wildlife areas in England. English Nature Research Reports No 182. Peterborough, UK.
Pugh, D. (2008). *Socio-economic indicators of marine-related Activities in the UK economy*. The Crown Estate, 68 pp. London, UK.
Pugh, D., & Skinner, L. (2002). A new analysis of marine-related activities in the UK economy with supporting science and technology. IACMST Information Document No. 10. Southampton Oceanography Centre, Southampton, UK: IACMST.
Pye, K., Saye, S. E., & Blott, S. J. (2007). Sand Dune processes and management for flood and coastal defence, Parts 1 to 5. Joint DEFRA/EA Flood and Coastal Erosion Risk Management R & D Programme, R & D technical report FD1 1302/TR. London, UK: DEFRA/EA.
Rees, S. E., Rodwell, L. D., Attrill, M. J., Austen, M. C., & Mangi, S. C. (2010). The value of marine biodiversity to the leisure and recreation industry and its application to marine spatial planning. *Marine Policy, 34*, 868–875.
Ressurreição, A., Gibbons, J., Dentinho, T. P., Kaiser, M., Santos, R. S., & Edwards-Jones, G. (2011). Economic valuation of species loss in the open sea. *Ecological Economics, 70*(4), 729–739.
Ressurreição, A., Gibbons, J., Kaiser, M., Dentinho, T. P., Zarzycki, T., Bentley, C., Austen, M., Burdon, D., Atkins, J., Santos, R. S., & Edwards-Jones, G. (2012). Different cultures, different values: the role of cultural variation in public's WTP for marine species conservation. *Biological Conservation, 145*(1), 148–159.

Ruiz-Frau, A., Hinz, H., Edwards-Jones, G., & Kaiser, M. J. (2013). Spatially explicit economic assessment of cultural ecosystem services: Non-extractive recreational uses of the coastal environment related to marine biodiversity. *Marine Policy, 38*, 90–98.

Salem, M. E., & Mercer, D. E. (2012). The economic value of mangroves: A meta-analysis. *Sustainability, 4*(3), 359–383.

Samonte-Tan, G. P., White, A. T., Tercero, M. A., Diviva, J., Tabara, E., & Caballes, C. (2007). Economic valuation of coastal and marine resources: Bohol Marine Triangle, Philippines. *Coastal Management, 35*(2–3), 319–338.

Saunders, J., Tinch, R., & Hull, S. (2010). *Valuing the marine estate and UK seas: An ecosystem services framework.* The Crown Estate, 54 pp. ISBN: 978-1-0-906410-15-5. London, UK.

Sen, A., Harwood, A., Bateman, I. J., Munday, P., Crowe, A., Brander, L., Raychaudhuri, J., Lovett, A. A., Foden, J., & Provins, A. (2014). Economic assessment of the recreational value of ecosystems; methodological development and national and local application. *Environment and Resource Economics, 57*(2), 233–249.

Shepherd, D., Burgess, D., Jickells, T., Andrews, J., Cave, R., Turner, R. K., Aldridge, J., Parker, E. R., & Young, E. (2007). Modelling the effects and economics of managed realignment on the cycling and storage of nutrients, carbon and sediments in the Blackwater estuary UK. *Estuarine, Coastal and Shelf Science, 73*(3), 355–367.

Smith, M. D., & Wilen, J. E. (2003). Economic impacts of marine reserves: The importance of spatial behavior. *Journal of Environmental Economics and Management, 46*(2), 183–206.

Stål, J., Paulsen, S., Pihl, L., Rönnbäck, P., Söderqvist, T., & Wennhage, H. (2008). Coastal habitat support to fish and fisheries in Sweden: Integrating ecosystem functions into fisheries management. *Ocean & Coastal Management, 51*(8), 594–600.

Tinch, R., & Ledoux, L. (2006). Economics of managed realignment in the UK. Environmental Futures Limited. Norwich, UK.

Turpie, J. K., Heydenrych, B. J., & Lamberth, S. J. (2003). Economic value of terrestrial and marine biodiversity in the Cape Floristic Region: Implications for defining effective and socially optimal conservation strategies. *Biological Conservation, 112*(1), 233–251.

Tuya, F., Haroun, R., & Espino, F. (2014). Economic assessment of ecosystem services: Monetary value of seagrass meadows for coastal fisheries. *Ocean & Coastal Management, 96*, 181–187.

UK National Ecosystem Assessment. (2011). *The UK National Ecosystem Assessment technical report*. Cambridge: UNEP-WCMC.

UKMMAS. (2010). *Charting Progress 2: Productive seas evidence group feeder report*. Published by Defra on behalf of UKMMAS.

Unsworth, R. K., Cullen, L. C., Pretty, J. N., Smith, D. J., & Bell, J. J. (2010). Economic and subsistence values of the standing stocks of seagrass fisheries: Potential benefits of no-fishing marine protected area management. *Ocean & Coastal Management, 53*(5), 218–224.

Van der Meulen, F., Bakker, T., & Houston, J. (2004). The costs of our coasts: Examples of dynamic dune management from Western Europe. *Coastal dunes* (pp. 259–277). Springer-Verlag Berlin Heidelberg.

Voke, M., Fairley, I., Willis, M., & Masters, I. (2013). Economic evaluation of the recreational value of the coastal environment in a marine renewables deployment area. *Ocean & Coastal Management, 78*, 77–87.

Wattage, P., Glenn, H., Mardle, S., Van Rensburg, T., Grehan, A., & Foley, N. (2011). Economic value of conserving deep-sea corals in Irish waters: A choice experiment study on marine protected areas. *Fisheries Research, 107*(1), 59–67.

Windle, J., & Rolfe, J. (2013). Estimating nonmarket values of Brisbane (State capital) residents for State based beach recreation. *Ocean & Coastal Management, 85*, 103–111.

Zheng, W., Shi, H., Chen, S., & Zhu, M. (2009). Benefit and cost analysis of mariculture based on ecosystem services. *Ecological Economics, 68*(6), 1626–1632.

# Chapter 7
# Scenarios Explored with Delphi

**Paul Tett and Laurence Mee**

## 7.1 Introduction

The purpose of this chapter is to demonstrate the use of mini-Delphi scenario workshops to consider how UK coastal and marine ecosystem services might change during the next half century, taking account of socio-political, as well as climate, change. Chapter 3 introduced numerical models as tools for predicting the future, and the work of the IPCC has shown how 'General Circulation Models' (GCMs) can be used to simulate climate change. Why not, then, couple a GCM to a model of a coastal social-ecological system?

There are several reasons why not. Firstly, because the discipline of social-ecological modelling is in no way as advanced as that of climate change modelling. Secondly, because complex systems include feedback loops that can amplify initial uncertainties and render prediction extremely imprecise. And finally, because complex numerical models are costly and time-consuming to set up and run.

In this chapter, we report an alternative to modelling, involving a 'Delphi' expert workshop. Our case study explored how ecosystem services might change under five socio-political scenarios, and how they might respond to social or ecological shocks.

## 7.2 Delphi

Unlike numerical computers, conscious human minds have very little algorithmic capacity; we use other ways of assessing evidence to make judgments in complex cases. Experts have in-depth knowledge of a particular domain but may be biased

P. Tett (✉) • L. Mee
Scottish Association for Marine Sciences (SAMS), Oban, Argyll PA37 1QA, UK
e-mail: Paul.Tett@sams.ac.uk



R.K. Turner, M. Schaafsma (eds.), *Coastal Zones Ecosystem Services*,
Studies in Ecological Economics 9, DOI 10.1007/978-3-319-17214-9_7

when extrapolating beyond this domain or into the future, and may be over-influenced by current disciplinary paradigms or a dominant personality during discussions. Conversely, group discussions may result in a bland, 'lowest-common denominator' outcome.

In classical times, the oracle at Delphi in Greece provided cryptic guidance for decision-makers. The modern 'Delphi method' was developed c. 1960 by the RAND corporation to improve expert-group judgements. Initially it involved interaction at a distance, by (i) the solicitation of anonymous responses to formal questionnaires, (ii) iteration with controlled feedback to participants, and (iii) an appropriate statistical method for aggregating opinions in the final round (Dalkey 1969; Donohoe 2011). Face-to-face discussion in a 'mini-Delphi' (Green et al. 2007) can shorten the process. Delphi methods have been recommended both for evaluating solutions to current problems when empirical evidence is lacking (e.g. Powell 2003), and for forecasting long-term developments (Cuhls 2001).

Models themselves depend on collective expertise and validation against past events to justify extrapolation to the future. Our contention is that expert workshops, run according to mini-Delphi principles, with opportunities to examine validity claims and re-assess initial assumptions, might provide a rough and ready estimate of future possibilities of equal reliability, but at much lower cost in cash and time, than may be obtained from complex social-ecological models. Of course, experts are not precluded from using model results, where available, as evidence.

## 7.3 Scenarios

The use of scenarios in planning and 'forward-looks' also began in the 1960s, and began to be applied to environmental matters towards the end of the century. The Millennium Ecosystem Assessment (MEA 2003) aimed to:

> use scenarios to summarize and communicate the diverse trajectories that the world's ecosystems may take in future decades. Scenarios are plausible alternative futures, each an example of what might happen under particular assumptions. They can be used as a systematic method for thinking creatively about complex, uncertain futures.

Whereas IPCC (2007, updated 2013) used a set of socio-political scenarios to generate schedules for future emissions of green-house gases and thus for predicting climate change, we took one climate schedule as a given and focussed on the potential consequences of several different socio-political scenarios for the use and sustainability of the UK's marine ecosystem services. The scenarios used in our case study were distinguished by differences in (i) the importance of market forces (versus other methods of resource allocation) and (ii) the dominant level of environmental government (from local through national to supranational).

The horizontal axis in Fig. 7.1 relates to personal dispositions to behave – at one extreme – as autonomous and competing individuals, interacting with others through bargaining, or – at the other extreme – as beings whose actions are mainly socially

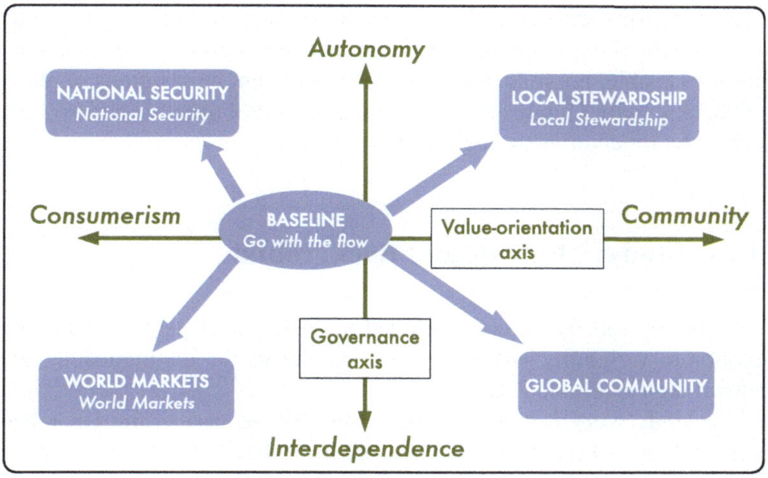


**Fig. 7.1** A psycho-socio-economic state space defined by two axes (relating to societal governance and individual dispositions towards consumerism or community) and containing the scenarios used in the case study. Based on UKCIP (2000), Pinnegar et al. (2006) and Cooper et al. (2008). Additional text (e.g. 'Go with the Flow') refers to the scenarios of Haines-Young et al. (2011) used in the first UK NEA

determined (Douglas 1970; Wildavsky 1987). In modern societies, in which money provides the main 'steering medium' (Habermas 1987), the first disposition – orientation to 'consumerism' – provides the basis for a society in which the satisfaction of well-being needs (and thus the distribution of resources) is largely dealt with by markets. The second disposition – orientation to 'community' – can underpin either hierarchical societies (in which power is the steering medium) or collectives without a formal power structure.

The second, vertical, axis relates to the large-scale institutions of modern societies, and the way in which these institutions operate across scales. We have called this axis 'governance' with the implication that it concerns institutions that 'steer' societies in relation to their geo-political environment. At the level of the nation state, these institutions include parliaments, central banks, legal systems, and armed forces. At higher levels there are organizations such as the EU, the UN, and the WTO. At lower levels there are local governments and 'civil society'. At the 'interdependence' end of the axis, global institutions control nation states, and – in the ultimate – citizens, through the 'steering media' of power or money. Or perhaps, by means of a global exchange of empathy and information via the world-wide-web. At the 'autonomy' end of the axis, the lower levels fully control operational and collective levels of governance. This might be a world in which states devolve most powers to localities, retaining mainly the constitutional level of governance: or, alternatively, a world made up of a thousand small polities, like the self-governing cities of ancient Greece.

Our five scenarios are located in this social-political state-space. One of them is 'Baseline', the forward projection of the current state of affairs. The others are supposed to be plausible, possible, and internally consistent descriptions of alternative states of society. Actual futures might combine several of these scenarios (i.e. might be described by trajectories through the state space).

## 7.4 Case Study: Methods and Background

A pilot study showed that account needed to be taken of geomorphological and socio-economic variability within Britain, leading to the distinguishing of three geo-political regions (Table 7.1).

The case study workshop was convened in 2013 as part of the UK's National Environmental Assessment Follow-On phase (NEA-FO). Twenty-six experts attended, invited on the basis of their knowledge of marine ecosystems and their services, and their willingness to engage in both role-playing and 'communicative action'. Communicative action, aimed at increasing mutual understanding of a topic, involves the making and hearing of 'discursively redeemable validity claims', and may be contrasted with strategic action aimed at achieving a successful outcome e.g. for the institution one represents (Habermas 1984). Role playing was necessary because participants were asked to briefly inhabit imagined worlds in which they act according to values other than their own. The aim was to combine both aspects, so that participants were able to evaluate and document outcomes (and check them for consistency) irrespective of their orientation towards those outcomes.

The participants included academics and stakeholders in environmental governmental organisations and NGOs; some also had expertise in workshop facilitation. The working methods were those of 24-h, 'mini-Delphi' process. The topic was that of changes in UK marine ecosystem services between 2013 and 2060 under the scenarios shown in Fig. 7.1. The scope of the exercise was defined as the UK's coastal and marine area but incorporating any necessary drivers beyond it. The time horizon of 2060 is within the timeframe of the UK Office of Budget Responsibility's Fiscal Sustainability Report projections (OBR 2012) for the next 50 years. A novel aspect of the workshop was consideration of the effect of shocks.

For simplicity, participants were provided with a single scenario for climate change, that predicted by IPCC in 2007 from the A1B greenhouse gas emissions schedule. UK coastal seas were expected to warm by 1–2 °C by 2060, to become slightly fresher (due to increased rainfall and runoff), and to remain stratified for a few days longer in each year (Jenkins et al. 2009). Mean sea level rise was taken from a high emissions scenario as 3 mm/year (totalling 0.2 m between 1990 and 2060 and with 50 % error bars; Lowe et al. 2009). Isostatic changes in land elevation would increase the relative mean rise to 0.3 m in the southeast and southwest and keep it at about 0.2 m in the north and north-west. To these small changes must be added the greater threat from storm surges. Although rare events, surges could, in the worst (simulated) case, combine with sea-level rise to add 1.5–2 m to present-

**Table 7.1** Geo-political subdivisions of Britain. Non-arable farming includes: sheep and dairy; use of moorlands for sports hunting; extensive conifer plantations. Isostatic changes (part of recovery from loading by ice-caps) are several mm/year (Bradley et al. 2009)

| Region | Socio-economic characteristics (2013) | Maritime characteristics | Major trends, 2013–2060 (in Baseline) |
|---|---|---|---|
| South-East England: lowlands overlying young (Mesozoic and Cenozoic) sedimentary rocks; dry warm climate; isostatically sinking | Highest population densities and wealth (albeit not uniformly distributed); arable farming and greatest use of fertilizers; the UK's largest international trading ports | Eroding sinking coastlines, shallow agriculturally polluted (nutrient-enriched) estuaries, mainly shallow, mixed, turbid coastal seas | Largest increase in population, real GDP and city development; decrease in agricultural production; increased household consumption (goods, energy) and demand for recreation; large increase in tourism and increase in services GVA |
| Wales and the North and West of England: hill country and valleys, overlying older (Palaeozoic) sedimentary and metamorphic rocks; cooler, wetter climate | Moderate population densities, with most clustered into impoverished post-industrial towns and cities; declining ports; non-arable farming; Welsh government has some marine environmental responsibilities | Mostly stable and scenically attractive coast (cliffs and beaches); some historically polluted and presently nutrient enriched estuaries and coastal waters; offshore seasonal thermal stratification | Moderate growth in population, real GDP and cities; increased energy production as well as consumption; increased recreational demands; some increase in tourism; increase in industrial GVA |
| Scotland: mountainous Highlands of old (preCambrian) granitic rocks, north of older sedimentary rocks; cooler wetter climate; isostatically rising except outer islands | Low population densities except in the central belt (Glasgow-Edinburgh); non-arable farming; devolved Scottish government with environment protection agencies and responsibility for many aspects of 61 % of the UK's EEZ | Many islands, firths and sea-lochs, with alternating exposed and sheltered waters; high runoff leading to frequent haline stratification; offshore waters with seasonal thermal stratification; the largest part of the UK's fishery and marine energy resources | Moderate growth in population, real GDP and cities; increased energy production as well as consumption; increased recreational demands; large increase in tourism; increase in services GVA; increased aquacultural production |

day astronomical high tide in parts of the west coast of Britain, and in East Anglia and the Thames estuary.

The five socio-political scenarios were those shown in Fig. 7.1 and listed in Table 7.2. The details (Cooper et al. 2008) provided to participants are given at the start of each subsection in Sect. 7.5. Scenarios 4–7 were discussed in subgroups of about 6 persons; shocks were discussed in physical-ecological and socio-economic subgroups, each of about 12 people; the baseline scenario was discussed (twice) in plenary.

Two sorts of data were obtained. Qualitative data took the form of narrative reports from subgroups, together with the comments recorded in the assessment forms. The reports were used to prepare the descriptive accounts for each scenarios. Inevitably, there was discussion concerning the desirability and feasibility of the world-views in the scenarios, as well as their implications for ecosystem services, and this is reflected in Sect. 7.5.

Participants were asked to use a Likert-type 5-point scale (Likert 1932; Clason and Dormody 1994) to assess the likely change in each service, in each geophysical region, under given socio-economic scenarios, assuming the pattern of climate change already described. Scores ranged from −2 (strong view of deterioration) to +2 (strong view of improvement). Scores were averaged over group participants, for each service and region, and expressed as whole numbers between −20 (unanimous strong view that service will worsen) and +20 (unanimous strong view that service will improve). Results are shown in Fig. 7.2, colour-coded from brown (worsening) through white (no change) to green (improving).

## 7.5 Case Study: Outcomes from Scenarios

### *7.5.1 Baseline Scenario*

This projects current trends in the existing state of UK society and economy. In addition to the socio-economic changes in Tables 7.1 and 7.2, the following were also assumed:

- UK Seas will be spatially planned and that projected activities (e.g. areas licensed for renewables development, Marine Conservation Zones, decommissioning of North Sea oil, expansion of oil and gas extraction in deeper waters, some Carbon Capture Schemes) will continue;
- Existing policies, mostly resulting from EU drivers such as the WFD and the MSFD, will be fully implemented (as a consequence of the UK Marine and Coastal Access Act 2009 and Marine (Scotland) Act 2010); there will be multiple iterations of the EU CFP, and increasing UK regional devolution.

There was mild optimism in the workshop about the sustainability of most services during the first round of scoring, which was tempered during the second round. A key reason for this optimism was the view that national and regional

**Table 7.2** Comparison of several sets of scenarios: NEAFO: scenarios used in this chapter; AFMEC (Alternative Future Scenarios for Marine Ecosystems): Pinnegar et al. (2006); ELME (European Lifestyles and Marine Ecosystems): Cooper et al. (2008); NEA 2011 (National Ecosystem Assessment): Haines-Young et al. (2011); MEA (Millennium Ecosystem Assessment): MEA (2005); UKCIP (UK Climate Impacts Project): UKCIP (2000)

| | Scenario in outline | NEA 2011 name | NEAFO name | Names in other work |
|---|---|---|---|---|
| 1 | Projection of present conditions and trends, including growth in population, real GDP, debt, energy consumption, service industries, tourism, transport, international trade, city and coastal development, decrease in fisheries catch/effort (see also Table 7.1) | Go with the Flow | Baseline | Business as Usual Baseline (ELME) Conventional Development (AFMEC, UKCIP) |
| 2 | National conservation funded from global markets | Green and Pleasant Land | | |
| 3 | Global free-market and environmental standards reconciled through valuing and nationally managing ecosystem services | Nature@Work | | TechnoGarden (MEA) |
| 4 | Strong subsidiarity, emphasis on environment and equity: 'green cantons' | Local Stewardship | Local Stewardship | Local Stewardship (AFMEC) Local Responsibility (ELME) Adapting Mosaic (MEA) |
| 5 | Strong state and protection of national market economy: 'patriotic individualism' | National Security | National Security | Fortress Britain (AFMEC) National Enterprise (ELME & UKCIP) Order from Strength (MEA) |
| 6 | global growth and free markets: 'competitive libertarian individualism and big companies' | World Markets | World Markets | World Markets (AFMEC, ELME, UKCIP) |
| 7 | globalization for equity and environment as well as markets : 'the Nordic social democratic model' | | Global Community | Global Commons (AFMEC) Global Community (ELME) Global Orchestration (MEA) Global Sustainability (UKCIP) |

| | Scenario | Baseline (1) | | | Baseline (2) | | | National Security | | | World Markets | | | Global Community | | | Local Stewardship | | |
|---|---|---|---|---|---|---|---|---|---|---|---|---|---|---|---|---|---|---|---|
| Final Ecosystem Service | Region | SEE | N&W &W | SC | SEE | N&W &W | SC | SEE | N&W &W | SC | SEE | N&W &W | SC | SEE | N&W &W | SC | SEE | N&W &W | SC |
| Wild fish/ shelfish/seaweed | | -1 | 1 | 5 | -5 | 1 | 6 | 1 | 1 | 1 | -18 | -18 | -13 | 20 | 20 | 20 | 12 | 13 | 13 |
| Cultured fish/ shelfish/seaweed | | 7 | 9 | 16 | 2 | 5 | 13 | -2 | 2 | 5 | 1 | 16 | 18 | 3 | 10 | 10 | 10 | 9 | 14 |
| Genetic resources | | 1 | 4 | 4 | -1 | 0 | 2 | -6 | 10 | -8 | -14 | -14 | -14 | 0 | 0 | 0 | 6 | 8 | 8 |
| Climate regulation | | 1 | 4 | 3 | 1 | 2 | 4 | -5 | 2 | -8 | -16 | -18 | -18 | 18 | 10 | 10 | -3 | -3 | -2 |
| Natural hazard protection | | 5 | 6 | 3 | 2 | 3 | 4 | -9 | -6 | -4 | -10 | 2 | -2 | 18 | 10 | 10 | 5 | 7 | 7 |
| Waste breakdown/ detoxification | | 2 | 4 | 4 | 3 | 4 | 4 | -9 | -7 | -5 | -12 | -4 | -4 | 1 | 10 | 10 | 5 | 8 | 5 |
| Meaningful places - Socially valued seascapes | | 3 | 8 | 7 | 2 | 6 | 9 | 8 | 5 | 7 | -18 | -5 | -9 | 1 | 10 | 10 | 8 | 8 | 8 |

**Fig. 7.2** Synthesis of scores for all scenarios (UK NEAFO 2014). The values show the average scores for each option, scaled to the number of participants registering an opinion in each case, so that +20 (colour-coded *green*) implies a consensus about a strong opinion that all components of the service will increase, and -20 (colour-coded *brown*) implies a consensus about a strong opinion that all will worsen. *SEE* South-East England, *N&W&W* North-west and south-west England and Wales, *SC* Scotland

environmental protection would become increasingly effective, supported by a public increasingly ready to accept proper costing of externalities. Differing regional trends were expected, as a result of lower population densities and greater recognition of the value of the environment (in itself and as a provider of services) in the north and west of Britain, in contrast to higher rates of population growth, urbanization, and economic development in the south and east.

## 7.5.2 *National Security Scenario*

This was described to participants as follows:

- **Values & Policy**: Individualistic, highly personal consumption, low taxes, market-based, but strong commitment to national culture and interests. Little concern for social equity or environmental protection. Sovereignty retained or taken back to national level. Externally, erosion of EU powers, and weakening of WTO links by protectionist measures.
- **Demography**: Little inward migration and relatively low birth rates, although UK age distribution balanced to some degree by diminished longevity. Migration to internal growth 'hot spots' and average household size stable, but with household numbers increasing more slowly than under Baseline.

- **Economy**: Priority of growth undermined by protectionist policies. Focus on meeting internal demand and on security of supply. Nevertheless external trade to obtain food, and export goods or services in exchange, would likely require at least bilateral agreements with trading partners. Considerable variation in regional development.

Subgroup participants expected the UK to take a strongly protectionist stance and withdraw from agreements and institutions that were seen as undermining its sovereignty. Thus it would leave the EU and revoke national transposition of the CFP, the MSFD, Birds and Habitats Directive, etc. Membership of OSPAR, ICES and the International Maritime Organisation would continue, and a complex series of bilateral agreements would be negotiated with neighbouring states. Much attention would be paid to self-reliance for energy supply (nuclear, coal and deeper sea and Falkland oil) and there would be increased spending on protecting borders and trade (from immigration and smuggling). Consequently, state support for welfare and environment would decrease. Innovation would be difficult to finance. The marine biotech industry would stagnate or emigrate. There would be strong protection of property rights, including marine property for which the Crown Estate would become the de-facto regulator. With increased domestic tourism, landscape values would be paramount (albeit threatened by weakened control of pollution). The renewables industry would shrink. Environmental protection and planning would be more reactive than proactive. Heritage and conservation charities and public bodies would likely have greater influence than environmental-protection agencies.

It was thought likely that fisheries management would go through cycles of boom and bust as bilateral agreements with neighbours proved ineffective and effort controls crumbled. The difficult financial situation might, however, eventually led to the removal of all subsidies and this, combined with fuel price hikes, would lead to bankruptcies and reduced fishing effort. Subsequent franchising of rights to fishing companies might then lead to improved stock management, with the franchisees reaching voluntary agreements with neighbours, even if effort exceeded the optimum for maximum sustainable yield. Aquaculture would only be further developed for the 'luxury goods and exports' market (mainly salmon) but warmer temperatures might cause the spread of Pacific oysters which could become popular with local prospectors.

Because sea defences would become increasingly expensive as sea level rose, only valuable assets (such as London's commercial district) would be properly protected; other coastal areas might be lost during locally catastrophic 'un-managed realignments'. Pollution control laws were expected to remain at about the same level as 2013, but with declining compliance because of lack of enforcement. There would be increasing problems with cumulative impacts. Feedback from recreational users through strong local councils and landowner associations might maintain protection for beaches and bathing waters.

Subgroup participants were generally pessimistic in their scoring of ecosystem services, expecting most to decline. The exceptions were fisheries, as discussed above, and socially valued landscapes, reflecting a greater pride in the national countryside and the increase in domestic tourism.

### 7.5.3 World Markets Scenario

This was described as follows:

- **Values & Policy**: Libertarian, techno-centric, materialist consumerism. Presumption in favour of market provision. Growth more important than social equity, with environmental policy limited to correction and support of the market. Increased global interdependence and governance, through WTO and multinational corporations. Corporate governance starts to displace national government. Policy determined at regional trading bloc and international level. Rapid enlargement of EU.
- **Demography**: UK population growth slows overall but migration increases to meet demand for labour and reduces proportion of older people. Growth uneven across regions. Smaller and more numerous households.
- **Economy**: Rapid UK and global growth, with dismantling of trade barriers increasing intra- and extra- EU trade. Service sector dominates others, with decline of agriculture and manufacturing. Benefits of growth spread to some extent through 'spill over' effects.

Participants concluded that outcomes depended on the ability of governing bodies to correct for externalities. It might be that an international body would successfully impose strong environmental regulation/certification, on the grounds that continued growth requires functioning ecosystems. Significant environmental degradation might take place before the wider community – including financiers and investors – realised that this degradation impacted on profit potential, and consequently put the business world behind greater regulation. A fundamental element of such regulation would be a working market for carbon. It was thought likely that most natural assets would be privatised and managed on the basis of property rights. Fish stocks, for example, might be managed by a global system of tradable quotas, very likely leading to greater consolidation of fleets and enhancement of profitability. The owners of these (now private) assets would have a direct incentive to use them and their supporting ecosystems sustainably.

Should UK and global society, however, prove too myopic to take this path, a failure to manage externalities could lead to 'mega-death'. The key driver of this would likely be climate change beyond that of the IPCC (2007) A1B case. Should this happen, large global shifts in population might occur as lands became regularly flooded or drought-stricken. The resulting pressures on remaining natural resources would lead to an increasing downward spiral in the most impacted countries and to international conflicts over scarce resources such as oil. The only brake on such a course of events would be the insurance market via increasing charges as risk increased. Within the UK, the south-east would likely be most detrimentally impacted.

### 7.5.4 Global Community Scenario

This was described as follows:

- **Values & Policy**: Communitarian, with internationalist values and increasing globalization of governance systems to deal with large-scale, interconnected, problems. Balancing of economic, social and environmental welfare, with preference for latter and acceptance of high tax levels. Policy co-ordinated at EU and international level, but implemented at local level. EU more centralised, with less regional autonomy, and slower expansion. Environmental policy expands across policy sectors and is prioritised. Powerful, green, WTO favours environmental protection in trade disputes.
- **Demography**: Low birth rates offset by migration to meet demand for labour, with some increase in average age but relatively static distribution. Household size declining slowly, and numbers grow at historic rates.
- **Economy**: Growth constrained by tax levels and social and environmental objectives. Shift to services is slower than in Baseline. Growth in intra- and extra-EU trade, but with some inhibition by 'footprint' concerns. Development evenly distributed across regions and classes, though with some transitional variations.

In this world the goals are 'strong' sustainability based on a 'slow' growth philosophy and practice. Subgroup participants foresaw emphasis on maintaining and/or improving overall wellbeing and the stock of wealth (i.e. discounted present value of a future consumption stream anchored to all four forms of capital – physical, human, natural and social). Population growth would be stabilised. The global economic system and network of interdependencies would be radically reformed. Remits of some international institutions would be re-orientated towards the 'slow' growth strategy. For example, the WTO could have its 'fair trade' brief expanded to include environmental sustainability concerns. Banks would have their retail and investment activities completely separated. A 'Tobin' tax would be in force internationally, constraining international speculation and its destabilisation of financial, energy, property and commodity markets. The World Bank and IMF would be assigned a stronger regulatory role in environmental as well as financial management. Natural capital and its contribution to 'wealth' would become part of the national/international income/wealth accounting practice.

Overall, a more extensive and interventionist regulatory regime would be in place, and a stricter and 'smarter' set of policy measures operating at the international and national scales. International environmental agreements would be negotiated and rigorously enforced; green-house gas emissions would be limited to meet a 2°–3° warming target; and the Law of the Sea Convention would be given strong legal 'teeth', alongside integrated coastal management (ICM) and other marine related governance. There would be a preference for fish over meat as a protein source, which, given limits on wild fisheries yield, would need the development of sustainable, probably multitrophic, aquaculture and the resolution of siting conflicts.

The UK would be following a 'green' growth strategy with an emphasis on innovation and investment in resource saving and recycling technologies, covering, energy, water, waste and other raw materials. Public transport would be favoured over private transport. Supply chains would be made short. Product differentiation and persuasive advertising would be discouraged. Resource exploitation would be constrained by the precautionary and 'polluter pays' principles, and risk minimisation rules would have precluded exploitation of 'fragile' areas such as the Arctic. Such areas would be zoned and kept clear of all activities except scientific research. There would be more 'soft engineering' of coastlines, generating more salty wetlands for coastal defence, wildlife, and carbon storage, less interference with sand dunes, and reduced cost for maintenance of 'hard' defences. All this would take place in the context of ICM, and with reduction in waste generation and discharge, would have led to improvements in water quality.

The state would intervene to try to redistribute income and wealth to reduce the gap between the top and the bottom of the income distribution, through progressive taxation and other fiscal means. Attitudinal and behavioural change would be evident across both civil society and the business communities. Social networks would be encouraging new social norms focused on reflexive citizenship and corporate responsibility and ethics, including greater appreciation of cultural and environmental assets. Thus, a global and national culture involving the maximisation of short term desires and profits would be replaced by a culture favouring longer term needs and 'average' rates of return. Fair compensation and equity would be adopted as principles to be applied in any significant resource conflict/trade–off contexts.

Although this is an attractive vision, a society that tried to move in this direction would likely encounter resistance, and there could be an initial flight of capital and service industries eschewing the new tax regimes. This would only be resolved if other countries joined the common institutions described in the text. However, the scenario itself is a possible and consistent configuration for a global or continental society, even if its state-space location would be hard to reach from present co-ordinates.

### *7.5.5 Local Stewardship Scenario*

This was described as follows:

- **Values & Policy**: Communitarian, co-operative self-reliance. High levels of public services funded by high local taxation. Strong emphasis on social equity and environmental protection at the local level. Local government replaces national and supra-national governance. EU becomes more diverse with regional autonomy and fragmented policy.
- **Demography**: Population size stable, but relatively low birth rates and increased public health provision increases average age. General migration away from cities, with household size increases and household number reductions.

- **Economy**: Slow growth, exacerbated by tax levels, with increases in smaller scale production. Trade greatly diminished, but with some preference for intra-EU over external trade. Growth more even across communities.

Participants thought that local stewardship would prove to be effective in promoting improved conservation of coastal and near-shore marine ecosystems and sustainable use of the resources they generate. However, Local Stewardship approaches are vulnerable to strong external forces beyond their control. For example, local community management of fisheries might encounter difficulties offshore, where communities do not have the resources to implement fisheries management measures and impose them on out-of-area exploiters. Thus, increased devolution and subsidiarity would be a need for an enabling and back-up framework provided by UK federal and EU legislation. For example, problems arising from decoupling of terrestrial, coastal and marine systems management could be overcome by the application of integrated EU Directives, as pioneered by River Basin Management in the Water Framework Directive of 2000.

There would likely be regional differences in the capacity and effectiveness of local stewardship for resolving regional and national ecosystem management issues. Regions such as Scotland may have increased capacity to expand coastal and nearshore production of marine based protein to help feed the more densely populated areas of England. Likewise, parts of England have the climate and soils that can produce enhanced yields of carbohydrates to help meet the needs of people in Scotland. However, given the differences in population pressures and differing economic foci of the human resources between regions, there would likely be differing interests in and ability to foster local stewardship. For example, Financial Services in London and the southeast of England currently dominate the UK economy. The Global Markets outlooks involved in these activities may counteract the effectiveness of local stewardship in improving the management of ecosystems and maintaining the quality and quantity of renewable resource flows.

The effect of these reservations (about the tension between local stewardship and the need for national, continental or global scale regulation) was reflected in a wide range of individual scoring. Nevertheless, the majority of participants were optimistic about outcomes under this scenario.

### 7.5.6 *Quantitative Analysis*

Two general points emerged clearly from the quantitative analysis in Fig. 7.2: first, there was consistency between the first and second assessments of the 'Baseline' scenario; and, second, some scenarios were thought to hold better prospects for services than others.

## 7.6 Case Study: Effect of Shocks

A shock is a short-term disturbance. In ecological terms, it corresponds to a pulse perturbation (Bender et al. 1984), and contrasts with a sustained or press perturbation. In scoping potential shocks, Pinnegar et al. (2006) remarked that not all change is gradual; it may happen suddenly as a result of what we call shocks in this chapter, or it may happen slowly as a result of a build-up of change within an ecosystem or an accumulation of pressures on the system. A shock might act as the 'final straw' that tips an ecosystem from one regime into another. In our analysis we treat slow disturbances in terms of scenarios (including the single scenario of global climate change), and shocks as temporary increases in the pressures on the marine ecosystems from outside their boundaries.

Several sorts of physical and ecological shocks were considered: a storm surge sufficient to overtop the Thames barrier with consequent pollution of the south-eastern North Sea; a 6 months period of reduced light and sea-surface heating resulting from a volcanic eruption on Iceland; blooms of an invasive species comparable to the ctenophore *Mnemiopsis* in the Black Sea; an extreme summer resulting in sub-thermocline de-oxygenation over large areas of coastal sea. In the group's view, most marine and coastal ecosystems would recover from such pulse disturbances within a few years. This resilience arises partly from the biological community and partly from the open and well-flushed nature of the seas around the UK.

It is possible that a shock might cause an ecosystem to shift from one regime to another, but it is sustained, press, disturbances that are thought more likely to bring about such change. Certain sorts of shock, such as flooding with salt-water, might have long-term consequences for the integrity of coastal freshwater wetlands and the services they provide. Other shocks might impact directly on certain services, for example on aquaculture, but their long-term impact would depend on their effect on the socio-economic rather than the ecological system: if for example the owners of fish-farms afflicted by jellyfish blooms decided to redeploy their capital elsewhere. Finally, such shocks, it was thought, might have ecologically beneficial effects if they changed human perceptions of the environment and thus drivers of change. For example it might be decided to accept flooded areas as part of managed realignment of the coast, so diminishing the 'coastal squeeze' which greatly weakens the ability of littoral and supra-littoral communities to move and adapt to sea level change.

Amongst the examples of political and economic shocks considered by Pinnegar et al. (2006) was the break-up of the Soviet Union, and in particular its effects on its Baltic states, which gained independence but at high economic cost. Our workshop discussed the possibility of a break-up of the European Union. However, complete dissolution seemed unlikely; more realistic possibilities included failure of some EU member states with greater centralisation. The break-up of the UK was another possible shock. In either case it was thought that there would be minimal long-term disturbance of ecosystem services from those expected under the Baseline scenario. The main threat was from weakened governmental oversight of environmental quality and the use of ecosystem services.

The economic shock considered was that of a recession more severe than that experienced by the UK since 2008, perhaps accompanied by collapse of state revenues, and lasting for a decade or more. The likelihood would be that an impoverished government could not afford to enforce statutory protections of the marine environment, and thus that there would be increasing press disturbances of marine ecosystems through over-exploitation of services, as envisaged under the 'National Security' scenario. On the other hand, such an economic shock might lead to a significant change in society, and perhaps to one of the two 'green' scenarios and a stable zero growth economy.

## 7.7 Discussion

The oracle at Delphi is remembered for its ambiguous pronouncements. In 560 BCE it told Croesus of Lydia that, if he made war on the Persians, he would destroy a mighty empire. According to Herotodus, in his 'Histories', Croesus went to war, but the outcome was defeat for the Lydian empire. It would have been better for Croesus to see the oracular response not as a conditional prediction but as a reminder to think about the consequences of war. Perhaps it is best to see our own Delphi exercise as much as an exploration of the intersection of three sets of issues as an attempt at conditional predictions of the future. The issues are: climate change; changes in governance and social values; the effects of these on marine ecosystems and on the services we take from them.

As Charles Dickens implied, when he wrote (in 'A Tale of two Cities') that 'it was the best of times; it was the worst of times', there is no state of society that is not a mixture of good and bad. Eleanor Ostrom and colleagues have argued that there are no panaceas for environmental problems, no single recipes for ways in which society should be organised so as to move towards sustainability (Ostrom 2007, 2009). Discussions of scenarios may be creative ways to identify particular solutions to environmental challenges, and some of these solutions might emerge in responses to scenarios that at first glance promise little for environmental sustainability. That is to say, it may be better to see the benefits of a scenario exercise as resulting as much from the process of debating the options as from any predictive outcome (Haines-Young et al. 2011). Debate not only clarifies issues; as a process of 'communicative action' (Habermas 1984) it can help motivate deeper engagement with the issues. In this respect, Pahl et al. (2014) emphasises the importance of the narrative component of scenarios.

Despite individual differences in scoring change in particular ecosystem services under a given scenario, there was a clear outcome from the workshop, shown in Fig. 7.2. Opinion was that the World Markets and, to a lesser extent, the National Security scenarios would likely lead to strong impairments in most marine and coastal ecosystem services, whereas the Global Community and Local Stewardship scenarios would lead to improvements. Explanations involved the priority given to environmental sustainability in the last two scenarios, the primacy of the market in World Markets, and the reactive and partial nature of governance in National Security. There was fair consistency amongst the two scorings of the Baseline scenario, and the median opinion was slightly optimistic for most services. The key explanatory factor in this case was the view that current environmental legislation, mostly transpositions of EU directives, would be fully implemented and enforced. A minority opinion expected economic drivers to prove stronger than the will to protect the environment. Regional differences were expected under all scenarios, typically the result of a gradient from the southeast of England (where population and consequent pressures are highest) to Scotland (with mostly lower pressures and an environment suitable for aquaculture).

The workshop outcomes suggest ways in which present and near-future management practices could be modified to improve sustainability of ecosystem services. Thus the 'Global Community' scenario points to the benefits of 'soft engineering' of coastlines and of multi-trophic aquaculture. Such technologies might be of immediate value as well as providing resilience against climate change, and managed re-alignment of coasts might increase carbon sinks through creating or restoring wetlands. As discussed in Chap. 2, purpose-specific models could be used to explore the costs and benefits of such solutions.

Both discussions concerning shocks to the Baseline scenario led to the conclusion (paraphrased from Nietzsche) that 'what does not destroy us, makes us strong'. Marine ecosystems were seen as resilient against pulsed physical or ecological disturbance, and the UK socio-economic system was seen as similarly resilient against foreseeable political or economic shocks. As already noted, our focus was mainly on the response of marine ecosystems to these shocks. An event such as Thames Barrier overtopping would be catastrophic for many citizens, and although the socio-economic system would recover, the costs might fall unequally across social groups, as occurred when New Orleans was flooded by hurricane Katrina in 2005 (Vigdor 2008). Our optimism about the resilience of the UK socio-economic system is based on a view of effective multi-tier governance (at local, regional and national levels). The market economy might be less resilient, due to 'just-in-time' supply chains and the possibility of bank collapse.

Green policies have generally had low priority for UK governments, and have typically been developed and implemented as a response to external circumstances. EU membership and the transposition of EU directives into UK law have brought about considerable improvements in nature conservation and environmental quality (Burns 2013). It was the shocks of the storm-surge flooding in 1953 that led to developments such as the Thames Barrier, and workshop participants took the optimistic view that the hypothetical overtopping of this barrier might in the long run lead to the development of greater ecological and societal resilience through more widespread and managed coastal re-alignment. Of course, this assumes that a flooding shock to the social system would lead to rational choices about the most effective means of coastal defence.

## References


Bender, A., Case, T. J., & Gilpin, M. E. (1984). Perturbation experiments in community ecology; theory and practice. *Ecology, 65*, 1–13.

Bradley, S. L., Milne, G. A., Teferle, F. N., Bingley, R. M., & Orliac, E. J. (2009). Glacial isostatic adjustment of the British Isles: New constraints from GPS measurements of crustal motion. *Geophysical Journal International, 178*(1), 14–22.

Burns, C. (2013). *The implications for UK environmental policy of a vote to exit the EU* (9 pp). York: Friends of the Earth.

Clason, D. L., & Dormody, T. J. (1994). Analyzing data measured by individual Likert-type items. *Journal of Agricultural Education, 34*(4), 31–35.

Cooper, P., Etherington, L., Bell, S., Farmer, A., & Williams, J. (2008). *Socio-economic scenarios of European development and integrated management of the marine environment* (Working paper series, 2008.08, 53 pp). Bath: University of Bath, School of Management.

Cuhls, K. (2001). Foresight with Delphi surveys in Japan. *Technology Analysis & Strategic Management, 13*(4), 555–569.

Dalkey, N. C. (1969). *The Delphi method: An experimental study of group opinion* (RM-5888-PR, ix + 79 pp). Santa Barbara: RAND Corporation.

Donohoe, H. M. (2011). A Delphi toolkit for ecotourism research. *Journal of Ecotourism, 10*(1), 1–20.

Douglas, M. (1970). *Natural symbols: Explorations in cosmology*. Harmondsworth: Penguin Press.

Green, K. C., Armstrong, J. S., & Graefe, A. (2007). Methods to elicit forecasts from groups: Delphi and prediction markets compared. *Foresight: The International Journal of Applied Forecasting, 8*, 17–20.

Habermas, J. (1984). *The theory of communicative action* (Reason and the rationalization of society, Vol. 1). Boston/Cambridge: Beacon Press/Polity Press.

Habermas, J. (1987). *The theory of communicative action* (Lifeworld and system: A critique of fundamentalist reason, Vol. 2). Boston/Cambridge: Beacon Press/Polity Press.

Haines-Young, R., Paterson, J., Potschin, M., Wilson, A., & Kass, G. (2011). Chapter 25: The UK NEA scenarios: Development of storylines and analysis of outcomes. In *The UK National Ecosystem Assessment technical report. UK National Ecosystem Assessment* (pp. 1196–1264). Cambridge: United Nations Environment Programme World Conservation Monitoring Centre (UNEP-WCMC).

IPCC. (2007). In Core Writing Team, R. K. Pachauri, & A. Reisinger (Eds.), *Climate change 2007: Synthesis report. Contribution of working groups I, II and III to the fourth assessment report of the intergovernmental panel on climate change* (104 pp). Geneva: IPCC.

IPCC. (2013). Summary for policymakers. In T. F. Stocker, D. Qin, G.-K. Plattner, M. Tignor, S. K. Allen, J. Boschung, A. Nauels, Y. Xia, V. Bex, & P. M. Midgley (Eds.), *Climate change 2013: The physical science basis. Contribution of working group I to the fifth assessment report of the Intergovernmental Panel on Climate Change*. Cambridge/New York: Cambridge University Press.

Jenkins, G. J., Murphy, J. M., Sexton, D. M. H., Lowe, J. A., Jones, P., & Kilsby, C. G. (2009). UK climate projections: Briefing report. Met Office Hadley Centre, Exeter, UK (Version 2, 2010). Exeter: UK Meteorological Office Hadley Centre.

Likert, R. (1932). A technique for the measurement of attitudes. *Archives of Psychology, 140*, 1–55.

Lowe, J. A., Howard, T. P., Pardaens, A., Tinker, J., Holt, J., Wakelin, S., Milne, G., Leake, J., Wolf, J., Horsburgh, K., Reeder, T., Jenkins, G., Ridley, J., Dye, S., & Bradley, S. (2009). UK Climate Projections science report: Marine and coastal projections. UKCP09 scientific reports, 95 pp. Exeter: UK Meteorological Office Hadley Centre.

Millennium Ecosystem Assessment. (2003). *Ecosystems and human well-being: A framework for assessment*. Washington, DC: Island Press.

Millennium Ecosystem Assessment. (2005). *Ecosystems and human well-being: Synthesis*. Washington, DC: Island Press.
Office of Budget Responsibility. (2012). *Fiscal sustainability report* (159 pp). Norwich: The Stationery Office.
Ostrom, E. (2007). A diagnostic approach for going beyond panaceas. *Proceedings of the National Academy of Science of the United States of America, 104*, 15181–15187.
Ostrom, E. (2009). Beyond markets and states: Polycentric governance of complex economic systems. In K. Grandin (Ed.), *The Nobel prizes 2009*. Stockholm: Nobel Foundation.
Pahl, S., Sheppard, S., Boomsma, C., & Groves, C. (2014). Perceptions of time in relation to climate change. *WIREs Climate Change*. doi:10.1002/wcc.272.
Pinnegar, J. K., Viner, D., Hadley, D., Dye, S., Harris, M., Berkhout, F., & Simpson. M. (2006). *Alternative future scenarios for marine ecosystems*. Technical report, 109 pp. Lowestoft: CEFAS.
Powell, C. (2003). The Delphi technique: Myths and realities. *Journal of Advanced Nursing, 41*(4), 376–382.
UKCIP. (2000). *Socio-economic scenarios for climate change impact assessment: A guide to their use in the UK Climate Impacts Programme* (13 + 127 pp). Oxford: UK Climate Impacts Programme.
UK National Ecosystem Assessment Follow-on. (2014). *The UK National Ecosystem Assessment Follow-on: Synthesis of the key findings*. Cambridge: UNEP-WCMC, LWEC.
Vigdor, J. (2008). The economic aftermath of Hurricane Katrina. *Journal of Economic Perspectives, 22*(4), 135–154.
Wildavsky, A. (1987). Choosing preferences by constructing institutions: A cultural theory of preference formation. *American Political Science Review, 81*(1), 3–21.

# Chapter 8
# A Review of Marine and Coastal Ecosystem Services Data and Tools to Incorporate This into Decision-Making

**Justine Saunders, Nicola Beaumont, Jonathan P. Atkins, Aisling Lannin, Dan Lear, Ece Ozdemiroglu, and Tavis Potts**

## 8.1 Introduction

One of the five guiding principles of the UK Government's 2005 Sustainable Development Strategy "Securing the Future" is to use sound science responsibly to underpin the activities of government departments and organisations. The principle promotes an evidence-based approach to decision-making and policy development

J. Saunders (✉)
Fugro EMU Limited, 1 Mill Court, The Sawmills, Durley, Southampton SO32 2EJ, England
e-mail: jus_at_home@yahoo.co.uk

N. Beaumont
Plymouth Marine Laboratory, Prospect Place, Plymouth, Devon PL1 3DH, England
e-mail: nijb@pml.ac.uk

J.P. Atkins
Hull University Business School, University of Hull, Hull HU6 7RX, UK
e-mail: J.P.Atkins@hull.ac.uk

A. Lannin
Marine Management Organisation, Lancaster House, Hampshire Court, Newcastle upon Tyne NE4 7YH, England
e-mail: aisling.lannin@marinemanagement.org.uk

D. Lear
Marine Biological Association, The Laboratory, Citadel Hill, Plymouth, Devon PL1 2PB, England
e-mail: dble@mba.ac.uk

E. Ozdemiroglu
EFTEC, 73-75 Mortimer Street, London W1W 7SQ, England
e-mail: Ece@eftec.co.uk

T. Potts
Department of Geography and Environment, The University of Aberdeen, Elphinstone Road, Aberdeen AB24 3UF, UK
e-mail: Tavis.Potts@abdn.ac.uk



R.K. Turner, M. Schaafsma (eds.), *Coastal Zones Ecosystem Services*, Studies in Ecological Economics 9, DOI 10.1007/978-3-319-17214-9_8

through each stage, from identifying and monitoring environmental issues, to the consideration of all available policy options, development of the most appropriate management response and subsequent evaluation of policy effectiveness.

It is critical therefore to invest in research for new data and to develop the ability to discriminate between evidence which is reliable and useful, and that which is not. These data need to be made publicly available to ensure transparency in the decision-making process, to encourage open and rigorous public debate and to avoid unnecessary duplication of research. Accessible and user-friendly tools and guidance are then needed to apply that data for decision-making in a consistent, relevant and rigorous approach (Scott et al. 2014).

Historically, marine policy, management and data collection have focussed primarily on environmental concerns. However, if policies are to be successful, both in terms of uptake and achieving their desired goals, it is increasingly accepted that they should also take into account social and economic factors (see Chap. 2, Defra 2011; United Nations 1987). Ecosystem services assessment aims to provide a more holistic view of the value that the marine environment provides to human beings and how this value may be impacted by policy decisions.

Ecosystem services research requires a large range of environmental, social and economic information to inform the various levels of the assessment framework (see Chap. 2). Environmental information is needed on the distribution and status of coastal and marine features (e.g. habitats and species, sea space, sea water, substratum) along with an understanding of the ecological processes that influence them in order to appreciate the flow of final services. In order to then assess human wellbeing derived from these final services (i.e. goods and benefits in Chap. 2) information is needed on the value of the marine environment to human wellbeing, and where and how this value is extracted.

As a consequence, in the commercial and academic arenas, the application of coastal and marine ecosystem services research to the understanding and management of the environment is a rapidly growing field (Liquete et al. 2013). However, the availability of data and tools to apply such thinking to decision-making is also key. The Marine Environmental Data and Information Network (MEDIN)[1] is an open partnership which promotes sharing of, and improved access to, marine data in the UK. The MEDIN datasets for natural science data are comprehensive and well developed, but deficient in terms of social and economic data.

The first part of this chapter focusses on the findings of a project carried out by several of this chapter's authors which developed and analysed a metadata[2] catalogue of relevant UK marine social and economic data (MMO and Marine Scotland 2012a[3]), hereafter referred to as "the catalogue". This includes data measuring the impacts on wellbeing as well as financial values or economic activity associated with uses of the coastal zone. Issues regarding the interpretation of socio-economic

[1] http://www.oceannet.org/

[2] Metadata are essentially data which describe the data, for example where and how the data were collected, the format and location of data etc.

[3] www.scotland.gov.uk/Resource/0041/00412950.xls

data are discussed, specifically including how to handle uncertainty within data and how to utilise qualitative data.

Secondly, the chapter provides a review of the tools that may facilitate the incorporation of ecosystem services data into decision-making. This is a summary of a more detailed project based in the UK (MMO and Marine Scotland 2012b) although many of the findings and recommendations for the future research agenda apply globally. Finally a series of recommendations is put forward for future research.

## 8.2 Detailed Analysis of Data Availability by Ecosystem Service Category

This section provides an overview of data availability by ecosystem category focussing on the final goods and benefits derived from the coastal and marine environment (see Table 8.1). There are a number of data gaps related not just to social and economic data, but also the availability of environmental data on which to interpret and apply economic data (for example, the abundance and distribution of species used for fish feed and fertiliser).
Given that the methodologies for the valuation of some goods and benefits are still in development, it is difficult to even assess what datasets might be required. For example, the economic value of the sea's ability to assimilate waste is complicated as it is partly dependent on demand (i.e. how much waste needs to be assimilated by the environment) and partly dependent on supply (i.e. the capacity of environmental processes and components to store, break down and regulate particular volumes, concentrations and types of waste). Datasets on the demand for waste management includes annual information on the location, volume, intensity, and monitoring of regulated discharge of wastewater and the disposal of hazardous and non-hazardous waste. Much of this information is managed by either the relevant environment agencies (EA, SEPA and NIEA) or Cefas (licensed disposal sites for dredged material). Datasets on the supply of natural waste management services is more disparate and relates to the distribution and quality of ecosystem components such as saltmarshes, benthic sediments and bacterio-plankton that store and breakdown contaminants and pollutants.

There are no spatial layers describing the distribution of educational marine resources as data sources are too disparate and would require significant time and budgets to collate; related values therefore remain a large data gap. It may however be particularly important at a local-scale, e.g. places such as Plymouth where marine research forms a large part of the local economy.

Furthermore, no single data-layer exists on the location of naturally-occurring coastal defence features although environmental data exist on the location of saltmarshes and shallow subtidal sandbanks, for example, which might enable such a layer to be produced. The economic value of such features has been estimated at national scale (Jones et al. 2011) and there are some more specific case studies

**Table 8.1** Gaps in evidence on ecosystem goods and benefits

| Service group | Final ecosystem service | Goods/benefits[a] | Status of information on value and distribution and issues |
|---|---|---|---|
| Provisioning Service | Fish, shellfish and other wild/farmed species etc. | Food: fish, shellfish, algae, *Salicornia* and other wild/farmed food | MMO and Marine Scotland Science, Fishermap in England, Scotmap in Scotland and FishMap Môn in Wales, ICES reporting, Cefas, Defra, Inshore Fisheries and Conservation Authorities (IFCAs), Seafish, also derived layers in CP2, MCZ regional projects, MSFD assessments |
| | | Fish feed (wild/farmed/bait) | Data available from Fish Producer Organisations and Seafish but economic value poorly known as difficult to separate landings from that for human consumption |
| | | Fertiliser and biofuels | Economic value poorly known and from disparate sources – see Jones et al. (2011) for overview |
| | Ornamental materials (shells) | Ornaments | Economic value unknown |
| | Aquaria materials (fish, seaweeds) | Aquaria | Economic value unknown |
| | Genetic resources | Medicines and blue biotechnology | Value unknown. See Jones et al. (2011) for overview |
| Regulating Service | Climate regulation | Healthy climate | Disparate sources of research on the scale of carbon sequestration and storage although unit values agreed |
| | Natural hazard protection | Sea defence<br>Prevention of coastal erosion | Economic value estimated: raw data exists, e.g. on the distribution of natural features that provide natural hazard protection to collate a spatial layer of values |
| | Clean water and sediments | Waste burial, removal and neutralisation | Quantities of waste stored and broken down has been estimated for some habitats or marine features but economic value poorly known for reasons given in text (see Austen et al. 2011) |

(continued)

**Table 8.1** (continued)

| Service group | Final ecosystem service | Goods/benefits[a] | Status of information on value and distribution and issues |
|---|---|---|---|
| Cultural Service | Places and seascapes | Opportunities for tourism and nature watching (including recreation) | Social and economic data includes participation and visitor numbers and the related economic turnover from the sector. There are 31 spatial layers that might be used to spatially allocate these values |
| | | Spiritual and cultural wellbeing | Economic value unknown |
| | | Aesthetic and inspirational benefits | Economic value unknown |
| | | Human health | Economic value unknown |
| | | Education, research, knowledge | Datasets highly disparate and incomplete (see below) |

[a]Dark grey = minimal social and economic or value data; Grey = some data, but insufficient to support impending policy needs; Unshaded = reasonable, but not perfect, data coverage

(see Chap. 10 and 11). There is also guidance documents on assessing the costs of flood risk (Environment Agency 2010) and a number of examples of the application of the guidance.

Defra is conducting a study due to finish in 2015 to provide valuation of regulating, provisioning and cultural benefits that would arise from targets set under Marine Strategy Framework Directive (MSFD). The UK NEA Follow-on Phase has conducted a series of valuation studies for MPAs as well as local studies on a wide range of services (Kenter et al. 2014). There are also a wide variety of EU funded studies including VECTORS,[4] ValMER,[5] and ODEMM.[6] Finally, information on blue biotechnology and medicines may be available now through online access to licences and university registers of research. All of this future information may inevitably be useful for marine planning purposes in understanding the wider economic value of marine plan areas.

[4]http://www.marine-vectors.eu/

[5]http://www.valmer.eu/

[6]http://www.liv.ac.uk/odemm/

## 8.3 An Overview and Analysis of Marine and Coastal Ecosystem Services Data

### 8.3.1 Overview by Data Type

Ecosystem services related studies recorded in the catalogue, included the benefits of natural and man-made coastal defences, marine protected areas and water quality. However, there were very few ecosystem service specific studies, and few of these were original primary valuation studies (see Chap. 6).

Data on some ecosystem services were covered by social and economic datasets in the catalogue under categories such as fisheries and education. However, it is important to note that these datasets do not mean that all management, policy and academic needs can be met. For example there are extensive and readily available fishery datasets but less consistent data to enable assessments on whether stocks are being fished sustainably.

There are few social datasets but these are generally very comprehensive, freely available, national datasets, such as those held by Office for National Statistics (ONS) and Economic and Social Data Service (ESDS) with internationally recognised protocols and standards for data collection and management. However, ONS data may be restricted to a particular spatial scale, e.g. local authority level, with a charge for the provision of more detailed economic data at smaller local scales. In addition, although the ONS and ESDS published datasets are very comprehensive they do not hold a lot of information specifically related to the coastal and marine environment. A number of complex assumptions are required in order to apply the data to marine policy and planning at the appropriate sectoral and geographical scales. For example, national data on tourism will not distinguish between that directly related to coastal and marine activities and other activities, and careful assumptions are often needed to utilise such data. The MMO, with steer from Marine Scotland, conducted a project to explore whether and how ONS data could be interpreted to be more applicable to marine planning (MMO 2014). Unpublished data accessible through the ONS service Nomis and directly from the ONS for a fee should provide much of the necessary information to support marine planning.

The review also collated datasets on the geographic location of social and economic activities and infrastructure, both marine and coastal. This locational and other supporting data (e.g. habitat and species data) can, for example, provide an understanding of the spatial distribution of values. Scotland's marine atlas holds the most up to date information on the location of activities in Scottish waters and the MMO have published a Master Data Register, an extensive list of locational datasets for England used in marine planning[7] The Crown Estate also provide access to their related activity data on the Marine Data Exchange.[8] The availability of social and economic data specific to Wales is less clear although some data are likely accessible

[7] https://www.gov.uk/government/publications/master-data-register

[8] www.marinedataexchange.co.uk/

through the Welsh Government's online statistics and research portal.[9] Marine social and economic data for Northern Ireland is not held in a central place and may only be accessible through each individual government department.

### 8.3.2 *Spatial Scale of Data*

The catalogue was primarily focussed on collecting national level datasets. However, exercises such as Strategic Environmental Assessment (SEA), marine planning and Marine Protected Area (MPA) projects have provided a number of regional datasets. It is worth noting that more local datasets may also be available from site specific Environmental Impact Assessments (EIA), Local Authority projects (particularly with respect to tourism and recreation) and local conservation group projects, for example, valuation studies of reserves managed by the Royal Society for the Protection of Birds.

Furthermore, the available data are not always at the appropriate spatial scale. Data are either (1) collated at a national level, making coastal and marine-specific and spatially-allocated assessments difficult or (2) collated at a local or site-specific level, often for a single activity, making it difficult to scale up for national level assessments. There are a number of initiatives trying to improve this evidence base: The Pembrokeshire Coastal Forum have mapped economic values to areas of coastal recreation and tourism in two pilot areas of Pembrokeshire[10] and the MMO have compiled spatial data specifically on coastal recreation activities[11] and looked at ways to better utilise existing tourism data (MMO 2014).

### 8.3.3 *Temporal Distribution of Data*

Temporal information regarding the datasets was poorly recorded in the metadata and therefore difficult to assess without delving further in to each individual dataset. Of the social and economic datasets 60 % of metadata provided time series information. Given multiple changes in organisations over time it can be difficult to trace historical data. A significant number of the social and economic datasets were collected as part of one-off projects which were funded to support marine management and influence future policy, for example Charting Progress 2 (CP2), Scotland's Marine Atlas, the National Ecosystem Assessment (NEA) and MPA projects. Even those projects that are updated regularly may be dependent on and vulnerable to funding from internal memberships, for example the British Marine Federation economic reports on recreational boating activity. Social datasets were particularly disparate and often held in individual project reports.

---

[9] http://wales.gov.uk/statistics-and-research/?lang=en

[10] http://www.walesactivitymapping.org.uk/economic-valuation/

[11] http://webarchive.nationalarchives.gov.uk/20140108121958/http:/www.marinemanagement.org.uk/evidence/1043.htm

### 8.3.4 Application/Utilisation of Data

As noted above, a large number of datasets were produced for specific applications in decision-making. However, explicit evidence of the direct application of the data in policy development and marine management can be unclear and specific information on how the data have been used may only be available from the supporting documents, if at all.

Integrating understanding of limitations and knowledge gaps as well as being clear about how data are used in marine management and policy could result in more effective and efficient research, contributing to shared understanding between regulatory and academic sectors. Increasingly, regulators now publish their decision making process and the evidence used to increase transparency. The MMO and Marine Scotland also prepare metadata catalogues for all planning evidence work and provide access to data used (where it is legal to do so) so that others can explore and re-use data.

Complementary work is currently underway within the MEDIN community to assign Digital Object Identifiers (DOI's) to datasets to improve the tracking of their use (Socha 2013; BODC Published Data Library[12]). The UK government has committed to increasing transparency of data generally through its open data strategy[13] and the development and promotion of the Open Government Licence (OGL) for public sector information. Additionally there are marine specific data sharing processes required to support the Marine Strategy Framework Directive (MSFD) and Common Fisheries Policy (CFP) commitments.

## 8.4 Interpretation of Ecosystem Services Data

The previous sections highlight the diversity in quality and sources of the social and economic datasets. The correct interpretation of these data is essential if they are to be used both efficiently and effectively. Two aspects of interpretation are covered in depth in this section: uncertainty, and utilisation of qualitative data. Other aspects of valuation are covered elsewhere in the book (see Chap. 4).

### 8.4.1 Addressing Uncertainty in Ecosystem Services Data

In general, information on uncertainties in social and economic data is poorly recorded in the metadata, for example, information on years when surveys were not carried out, or information that the current dataset is undergoing review and is soon to be updated. However, such information is critical in the interpretation of the data.

---

[12] https://www.bodc.ac.uk/data/published_data_library/

[13] www.data.gov.uk/

With respect to ecosystem services there are many sources of uncertainty: in the distribution and quality of ecosystem components, understanding of economic value and the links between changes in ecosystem function or extent and consequent changes in service flows/benefits. Analysis of ecosystem services (whether for 'financial', 'market' or 'non-market' impacts) involves three steps:

(i) qualitative analysis that identifies the links between the changes in ecosystem function/extent and consequent changes in service flows/benefits
(ii) quantitative analysis that produces biophysical and other data about this linkage
(iii) economic analysis (market and/or non-market) that also takes account of the social and economic characteristics of the affected human population.

There are gaps and uncertainties in each of these three steps. For example in the first step, we may not know the link between the change in the ecosystem and the change in the associated services.

The knowledge gaps in step (i) naturally continue in step (ii) as an unknown impact cannot be quantified. Even for known impacts, data may not exist, or data may exist in one location for one change context but may need to be 'transferred' to other locations/contexts. This transference introduces uncertainty, the scale of which would depend on the similarities between the two locations or contexts (see Chap. 4 and 10).

Step (iii) could suffer from three types of uncertainty. First, the value data may not exist or may not be specific to the location or context of interest. Second, the value data may exist but may not be robust (e.g. survey data may come from very small samples and may be for very specific changes that limit the transferability of the results). Third, the value data may exist and be robust but is often incomplete. It is rarely possible, even when primary valuation research is commissioned, to have monetary value estimates for each type and scale of ecosystem service change. Therefore, the third step of the analysis, the appraisal, tends to produce results expressed in monetary and non-monetary units. The methods with which non-monetary estimates are obtained and the extent to which they are included in marine management, decision making and policy appraisal also contribute to the uncertainty surrounding the results. Integration of research about valuing the environment (HM Government 2011a, 2013; Saunders et al. 2010a) into analysis and decision making should reduce this type of uncertainty.

A final type of uncertainty stems from the scope of analysis. In most social and economic analyses (in all three steps) to date, the focus is usually on one pressure that causes ecosystem changes. If the same resource or service is subject to multiple on-going pressures, or to combinations of threats (natural or human induced), then an analysis focusing the baseline assessment on just one pressure could miss the dangers associated with the overall impacts. For example, when determining the impacts of aggregates extraction on fisheries it may be necessary to consider not only the direct impacts of extraction on fish habitats, but the bigger picture of threats facing fish populations, including overfishing, climate change and the availability of alternative habitats. Not doing so will lead to uncertainty in the results. Similarly most analyses do not take note of 'cumulative' impacts over time.

Factors that lead to uncertainty in individual economic value estimates include: the size of the impact being valued; the permanence of the impact; the affected (human) population; the valuation function; the timescale over which valuation occurs; and the discount rate (see Chap. 4).

#### 8.4.1.1 Handling Uncertainty in the Decision Making Context

Uncertainty (both in terms of gaps in data and uncertainty around the existing data) is inevitable. How much of an obstacle uncertainty is to efficient decision-making depends on (i) how easy it is to handle uncertainty in the decision making context; and (ii) how robust the data need to be for the purposes of a given decision-making context. To increase transparency and trust in decision-making, it is crucial to make clear what the evidence tells us, and what the uncertainties are. For this, the meta-data of the data used (or available) should be presented.

The level of uncertainty around an estimate can be gauged by undertaking sensitivity analysis, which involves re-estimating the economic analysis (e.g. CBA) using of a range of values for the parameters that represents a range around the true value. This is done by varying one parameter at a time to assess the effect on the result, which may be most appropriate when there are only a small number of parameters of concern. It is important to test as many assumptions as possible to see how different options fare in each run and what the net outcomes are.

Once the level of uncertainty has been determined, there are several methods available for clarifying the potential implications of this uncertainty for a given decision making context. The aim of these methods is to assist the marine manager or policy maker in understanding what uncertainty means in a specific context, and how to minimise the potential negative implications of this uncertainty. This requires good communication between the manager or decision/policy maker and the data provider. In addition, a number of more technical methods are available, including 'minimax' which selects between different options, having run the analysis under several different assumptions, so that the regret of making a wrong choice (selecting the wrong option) is minimised. 'Regret' in this context is defined as the difference between the net present value of the chosen option and that of other options.

When there are multiple parameters, each with significant uncertainties, Monte Carlo Analysis allows for these uncertainties to be assessed simultaneously in a single procedure. Monte Carlo Analysis relies on repeated random sampling to compute uncertainty estimates. Over a given domain of possible values (the range of values and the probability distribution across this range), repeated iterations generate a best estimate for the valuation and a confidence interval within which this value is expected to lie. In certain distributions (e.g. normal or triangular) the probability assigned to extreme values is very small, but their inclusion is necessary to ensure a true representation of the degree of uncertainty surrounding the estimate.

A strategic approach to use, especially when probabilities are not known, is the switching analysis. This helps answer the question 'How wrong does the analysis need to be for the results (the selected option) to be different, in other words, for the positive net present value for an option to become negative?' The switching value (in percentage) for benefits can be estimated simply by deducting the present value of benefits from the present value of costs and dividing the resulting amount by present value of benefits, and vice versa for switching value for costs. The higher the percentage value the less influence an individual (an uncertain or absent) factor has on the results. Note that this analysis does not change the level of certainty in the results, but presents the 'comfort zone' around them.

#### 8.4.1.2 How Much Uncertainty Is Acceptable?

In terms of how much uncertainty is acceptable, there is no theoretical benchmark against which a given uncertainty level can be judged. For risk information, when the probability and/or magnitude of change is known, the smaller the probability is the better. For uncertainty, this is left to how risk averse the decision context can afford to be; one way for decision makers to assess this are the costs of making a wrong decision (see above). It also depends on the phase of the policy or decision-making cycle in which the assessment is carried out (Brouwer 2008). Risk and coping with uncertainty guidance is provided in the UK Green Book (HM Treasury 2011).

There is no method of reducing uncertainty to zero, and as a result it is important to ensure stakeholder engagement and ownership of a decision, together with all its uncertainties. This can only be achieved if stakeholders are involved in a decision from the start and work together to address uncertainty and its impacts (e.g. commission new research, agree on assumptions to be tested in sensitivity analysis etc.) rather than use uncertainty as a reason for disagreement.

If the level of uncertainty is considered unacceptable, a valid option should be to delay the decision making until further relevant data can be collected. In Wales, for example, the recommendations for a network of Marine Conservation Zones (MCZs) have been delayed until more is understood about the potential ecosystem benefits and economic impacts (see Chap. 9). If the decision cannot be further delayed, the reasons for progressing despite uncertainty should be clearly communicated to all parties. In practical terms, economic valuation and CBA deal with risk (i.e. where probabilities are known) reasonably well, and with ambiguity (known outcomes, unknown probabilities) to some extent, through calculation of expected values and various forms of sensitivity analysis. However, economic methods are more limited where possible outcomes are unknown (see Chap. 2). In these cases, concepts of 'safe minimum standards' can be used for aspects of the natural environment that need to be safeguarded because they have critical functions that cannot be substituted or they are near critical limits (Barbier et al. 1990). Unfortunately in most cases it is not possible to identify such limits, and in some cases a highly precautionary approach may not be acceptable due to a disproportionate cost limitation on development.

### *8.4.2 Utilisation of Qualitative Data*

Qualitative data play a unique role in environmental management, but are often misused, misinterpreted or taken out of the context. Qualitative data does exist[14] and is accessible, but their use and meaning differ substantially from that traditionally collected for economic assessments.

How sectors, communities, or individuals respond to valuation, in the complex real life context of social, cultural and political action is at the heart of qualitative research. Detailed studies on valuation in a social context can highlight the potential consequences of implementation of an ecosystem service style policy instrument e.g. payments for ecosystem services, green taxes or positive subsidies. How valuations can actually be utilised and what their full consequences are can only be assessed by deliberative means, based on contextual analysis of the relevant political, social, and economic system, its networks and its stakeholders (see Chap. 2). These questions are likely to be controversial and result in public debate, resulting in winners and losers, and shifts in the allocation of resources.

While qualitative research is limited in terms of informing a measurable value for a particular ecosystem service (and often contests the utility of such approaches), its benefit lies in understanding how society responds to a particular policy instrument or scenario. Qualitative approaches may provide insight into the deeper and significant social values that are attached to seascape which are critical for political negotiation and implementation of policy.

A number of studies have explored 'intangible' values of communities towards the sea (e.g. Potts et al. 2011, 2012, 2014; Gee and Burkhard 2010). These data show that societal perspectives that emphasise aesthetic as well as practical aspects of the seas and non-market ecosystem services (e.g. climate regulation and scenery) are rated as important as economic maritime activities. However, as expected, countries vary in their views on particular services with significant differences in interpretation of relative importance, highlighting that analyses should consider the national and cultural context. Such qualitative data illustrates that coastal and marine values captured through economic metrics are not necessarily those of most importance to individuals and there is a clear research challenge in including some of the less easily quantified aspects of the marine environment in planning and decision making.

Data on non-market values around cultural ecosystem services are a case in point in terms of a lack of clear methodological guidelines, definitions or applications (Potts et al. 2014). There is a substantial need to expand, standardise and improve qualitative approaches so that they can complement and augment quantitative estimates.

Uncertainty surrounds the qualitative methodological processes used to determine values (methodological uncertainty) which are themselves subject to considerable debate. Furthermore qualitative data underpinning cultural ecosystem services are inherently context dependent, rich, and non-transferable to the broader

[14] For example the Ecosystem Services Indicators Database produced by the World Resources Institute holds both qualitative and quantitative data on ES indicators including examples of aesthetic and spiritual indicators. See: http://www.esindicators.org/

context. Caution should be used in extrapolation of qualitative datasets that cover deliberations over the meaning of place, power and context, and, while critical for understanding the application of the ecosystem services approach, contribute more to theoretical development rather than longitudinal assessment.

Including qualitative data in a metadatabase, such as the one explored here, is fraught with difficulty as qualitative data often do not fit into the set keywords and categories. However, this information is of key importance and further efforts must be made to describe datasets of a qualitative nature.

## 8.5 Tools

This section provides a general overview of the different types of tools (as part of the DSS, see Chap. 2), their functions and application throughout the planning process, their various strengths and weaknesses and data requirements.

Tools have a number of different functions throughout the marine policy development and licensing process including:

- Understanding the problem that needs management;
- Data mapping and visualisation;
- Development of policy or development options;
- Selection of sites to meet policy or development objectives;
- Assessment of the economic and social impacts of policy and development options;
- Monitoring and evaluation of policy objectives, targets and licensing conditions.

A summary of the various processes that might be involved under each function is provided in Table 8.2. Examples of applications are given from a regulatory perspective under marine planning and a developer's perspective under marine project development. Some of the tool functions are dependent on each other, for example, development of policy options (and underlying policy objectives) requires an understanding of the issue that needs managing, and site selection tools require mapping and visualisation routines. Processes involved in the development of plan and project options and Impact Assessments should aim to identify objectives, key issues and useful indicators to assist in the development of monitoring programmes. It is also worth noting that some of these tools are required throughout the decision-making process, for example, tools for mapping and visualising data will be important throughout the planning process for communication and stakeholder engagement.

The tool functions may involve both spatial and temporal models and to various degrees may help to:

- Incorporate data from ecological, economic, and social systems;
- Clearly assess management alternatives and trade-offs;
- Facilitate stakeholder participation and collaboration, community outreach and engagement;
- Evaluate progress towards management objectives.

**Table 8.2** A summary of the processes involved for each tool function and relevant planning and licensing applications

| | | **Examples in marine and coastal management** | |
|---|---|---|---|
| Tool function | Processes involved | Marine planning stages – regulator led aspects | Marine project stages – developer led aspects |
| Understanding the problem | Identify issues, constraints and future conditions, baseline assessment (dynamic and spatial) | Before planning starts – assessing why a plan is needed | What is the need for the project? Is an EIA required? |
| Data mapping and visualisation | Gather metadata and data, QA data, identify confidence levels, map data, make data available | Stakeholder engagement, Ongoing plan communication | Stakeholder engagement, Ongoing project communication |
| Development of options | Define objectives, explore scenarios, develop alternative management measures or project options | Plan development | Project development including early feasibility studies |
| Site selection tools | Define site criteria, explore scenarios, develop alternative spatial configurations, resolve spatial constraints | Refine any spatial aspects of the plan | Refine any spatial aspects of the project |
| Impact assessment | Evaluate the costs and benefits, strengths and weaknesses of baseline and alternative options | Sustainability Appraisal of the plan | EIA of the project |
| Monitoring and assessment | Gather monitoring data, assess performance indicators and evaluate plan/project | Plan monitoring and review | Post-consent monitoring and review |

Table 8.3 provides examples of the tools that are available under their various functions. The focus is on tools available in the UK although the review also extended to products available internationally. Links to all of the tools listed can be sourced from the report (MMO and Marine Scotland 2012b).

Overall, there are a large number of tools available for mapping and visualising data ranging from published reports to web-based maps that allow users to manipulate existing data and stakeholders to add their own data. Most of the government-led online web tools are secured with long-term funding to ensure that the initial investment in developing the tool and the effort made by stakeholders to populate it with data are protected.

A number of tools exist to help policy-makers identify likely policy and plan options from simple mapping exercises to complex and data-intensive simulation models. They are often useful in engaging with stakeholders and exploring the initial outcomes of different management options. There are few examples of such tools being

**Table 8.3** Summary of tools

| Function | General methods | Example products |
|---|---|---|
| Data mapping and visualisation | Web maps<br>Data catalogues<br>Reports | The MMO Marine Planning Portal<br>The Marine Conservation Zone portal<br>National Marine Plan Interactive<br>Marine Scotland Interactive<br>MaRS<br>EVRI (The Environmental Valuation Reference Inventory) |
| Development of options | Virtual and real world simulations<br>Mapping<br>Modelling of outcomes<br>Initial coarse-level impact assessment | CoastRanger<br>Co$ting Nature<br>Coastal Resilience<br>Integrated Valuation of Ecosystem Services and Trade-offs (InVEST)<br>Multiscale Integrated Models of Ecosystem Services (MIMES)<br>ARtificial Intelligence for Ecosystem Services (ARIES) |
| Site selection | Geo-spatial modelling<br>Cost optimisation models<br>Mapping of constraints | MaRS<br>Touch-table<br>Marxan<br>Multipurpose Marine Cadastre (MMC)<br>MarineMap |
| Impact Assessment | Cost-Benefit Analysis (CBA)<br>Cost-Effectiveness Analysis (CEA)<br>Multi-Criteria Assessment (MCA)<br>Trade-off Analysis<br>Life-Cycle Analysis (LCA)<br>Bioeconomic models<br>Risk Assessment | DEFINITE (decisions on a finite set of alternatives)<br>IMPLAN (IMpact analysis for PLANning)<br>InVitro (www.cmar.csiro.au/research/mse/invitro.htm)<br>SolVES (Social Values for Ecosystem Services) (http://solves.cr.usgs.gov)<br>EMDS (Ecosystem-based Management Decision Support)<br>Cumulative Impacts model (www.nceas.ucsb.edu/globalmarine)<br>SPICOSA, ARIES, MIMES, InVEST |
| Monitoring and Assessment | | Marine Integrated Decision Analysis System (MIDAS)<br>Ecosystem Assessment and Reporting Tool (EAR) |

used in the licensing process, although they could be used to develop alternatives to the proposed project for consideration in the Environmental Statement.

The tools explored in this review ranged from simple mapping exercises to complex environmental and economic models. The level of complexity required from a tool is generally dependent on the level of risk involved in the decision making process (encompassing social, economic and environmental factors), the spatial scale of a plan or project (i.e. ranging from the development of large zonal wind farms to small coastal marine works) and the temporal scales being considered (ranging from short term to long term projects).

Site selection tools specifically facilitate the development of spatial options to address policy issues. They have had limited application in planning initiatives in the UK to facilitate engagement with stakeholders, because the assumptions used for assigning constraint criteria or cost optimisation models are not always easy to communicate. However, site optimisation models can be useful in engaging with stakeholders and facilitating the design of developments. Good practice guidance could be developed to assist the future application of such tools.

The largest functional area of tool proliferation is in assessing the impact of policies. There are at least seven general methods that are closely related to each other but vary slightly in the level of detail required, the involvement of stakeholders in the process and in the criteria being assessed. They can be broadly compared as follows (see also Chap. 2):

- CBA compares different policy options according to a monetary analysis that may include values of ecosystem services;
- CEA compares policy options according to an objective or outcome unit;
- MCA compares policy options according to scores and weightings rather than monetary units;
- LCA compares policy options according to alternative measures such as energy use or carbon emissions (monetary analysis may later be applied);
- Risk assessment compares policy options according to levels of acceptable risk;
- Bio-economic models compare policy options according to modelled interactions between the environment and human activities; and
- Trade-off analysis compares policy options according to stakeholder consensus often including one or several of the methods above.

Given the number of general approaches to impact assessment, it is not surprising that numerous products have been developed to facilitate decision-making. However, as part of the review, an email and telephone survey of University institutions specialising in marine impact assessment models indicated that there are few products developed specifically for application to the marine environment in the UK, and even fewer that incorporate ecosystem services.

Despite the lack of specific products, there have been several applied projects to provide spatial interpretation of economic values from the marine environment in order to understand the distribution of value throughout the UK and to inform impact assessments of policy options. These projects include CP2 (UKMMAS 2010), the UK NEA (2011), Valuing Change in UK Seas (Saunders et al. 2010b) and the recent baseline developed to inform the Impact Assessment for the Marine Strategy Framework Directive (Defra and Marine Scotland 2012). These projects have made a number of advances in understanding the assumptions and methodologies for spatially presenting economic values that have been agreed at a high level.

Very few of the tools are used to fully investigate options which ensure the sustainable development of the marine environment. For example, there are generally good figures on turnover or GVA for economic activities such as commercial fisheries, but these figures do not adequately capture the flow of economic stocks and whether this return is sustainable. The ability to do this would require bio-economic modelling of each individual managed stock along with indicators of Maximum Sustainable Yield (MSY) and Sustainable Stock Biomass (SSB). This information exists to varying levels of confidence, but agreed methods for the use this raw data to prepare spatially allocated data layers of the "sustainable" economic value of an activity have not yet been explored.

The review has highlighted that limited tools are available for assessing effectiveness of policies and further work should be focused towards tools that help to monitor and assess the achievement of policies, plans or projects and their objectives.

As explained in Sect. 8.4.1, it is important that tools are able to account for uncertainty where the quality of input data or understanding of relationships may be poor, e.g. through (spatially explicit) confidence assessments and sensitivity analyses. A few of the impact assessment tools are known to have this added functionality including MaRS, DEFINITE and EMDS.

The tools explored encompassed a range of data themes, including financial values associated with marine activities and their geographical location. Less common, particularly in the UK, was the inclusion in decision-making tools of indirect economic values (e.g. supply chain data and employment) and social data on the characteristics of coastal communities. This is partly due to the difficulties in geo-referencing data for the marine environment that have been collected according to terrestrial geographies.

## 8.6 Summary and Recommendations

### *8.6.1 Data*

(i) Data gaps: There is a clear management, policy and academic need for high quality UK-specific marine social and economic data[15] (HM Government 2011b; Liquete et al. 2013). The data available do not necessarily meet all the management, policy and academic requirements. The weaknesses highlighted above should help steer future research requirements. With regard to management there is a need for policy makers, regulators and their advisors to provide more detailed guidance regarding what data they require, and better integration between data collection initiatives and regulatory needs.

(ii) Spatial scale: The catalogue focused on datasets at the national level. However, local information may be held by local councils, site-specific academic research, project-specific environmental impact assessment and monitoring and studies by local conservation groups.

(iii) Links with natural science: Some of the data gaps in ecosystem services are present due to a lack of natural science data and unless we can quantify the changes in the natural environment we cannot value those changes. For example, initial Impact Assessments for MCZs (Defra et al. 2012) were unable to

[15] www.marinemanagement.org.uk/about/documents/strategic-evidence-plan.pdf

quantify beneficial ecosystem service changes without an understanding of the expected environmental change due to MCZ protection measures. The development of a data strategy for the coastal and marine environment should be undertaken with integrated ecosystem serv ices requirements in mind.

(iv) Qualitative data are also required particularly in the field of cultural ecosystem services. Including qualitative data in management, policy and decision making will require further efforts to describe, categorise and publish datasets of a qualitative nature, but could greatly improve the benefits for society.

(v) Central repository: The lack of a dynamic central repository for metadata hinders its usage as datasets are spread across a disparate range of sources making it difficult for any potential user to find the relevant datasets. This is being addressed in the UK through the development of a central MEDIN portal and the encouragement of more open access agreements including the Open Government Licence. This will ensure that the UK is implementing legislation across departments based on a standard social and economic evidence base that can be used with confidence and is regularly reviewed.

(vi) Standardisation of data management approaches: To ensure the data are robust, transparent and defendable, best practice must be applied throughout the data life-cycle, from creation to long-term curation. Such best practice includes the creation of data to meet open, stable, internationally agreed standards and formats. Generating data in this way is critical to facilitate the greatest degree of interoperability and reuse, and to maximise the data's value. To improve accessibility of data a single standard could be promoted and its importance communicated to researchers, managers and policy makers.

(vii) The study reported poor recording of metadata particularly relating to temporal aspects, the standards and protocols applied. The metadata and associated publishing protocols must be given an equal priority, to enable a high degree of visibility to the data, enabling validation and wider utilisation of the data.

### 8.6.2 Tools

The following recommendations were made to the MMO and Marine Scotland following the review of tools to apply social and economic data to decision-making:

(i) Explore the use of models to develop realistic alternative options for marine planning where this may in turn help to inform the licensing process.

(ii) Develop good practice guidance on the application of site optimisation models in consultation exercises with feedback gathered from stakeholders on the suitability of different site selection tools.

(iii) Investigate the feasibility of adapting existing impact assessment models and guidance for specific use in coastal and marine policy development and incorporating assessments of ecosystem services and human wellbeing.

(iv) Agree methodologies to provide spatial understanding of economic values through targeted workshops with government economists and industry groups.

(v) Support further research towards tools to help monitor and assess the achievement of policies, plans or projects and their objectives, such as the development of anthropogenic pressure benchmarks and associated spatial data layers.
(vi) Support further research to incorporate information on ecosystem services in planning tools.
(vii) Explore new methodologies and tools to better assess the sustainability of marine activities in the coastal and marine environment.

## References

Austen, M., Malcolm, S., Frost, M., Hattam, C., Mangi, S., Mieszkowska, N., Stentford, G., Burrows, M., Butenschön, M., Merino, G., Miles, A., & Smyth, T. (2011). *Chapter 12: Marine*. UK National Ecosystem Assessment, UNEP-WCMC, Cambridge.

Barbier, E. B., Markandya, A., & Pearce, D. W. (1990). Environmental sustainability and cost-benefit analysis. *Environment and Planning A, 22*(9), 1259–1266.

Brouwer, R. (2008). The role of stated preference methods in the Water Framework Directive to assess disproportionate costs. *Journal of Environmental Planning and Management, 51*(5), 597–614.

Defra. (2011). Mainstreaming sustainable development. Available at: http://sd.defra.gov.uk/documents/mainstreaming-sustainable-development.pdf

Defra, & Marine Scotland. (2012). Business as usual projections of the marine environment, to inform the UK implementation of the Marine Strategy Framework Directive.

Defra, The Joint Nature Conservation Committee (JNCC), & Natural England. (2012). Impact Assessment of designation of Marine Conservation Zones in England. Available at: http://publications.naturalengland.org.uk/publication/2071071

Environment Agency. (2010). Flood and Coastal Erosion Risk Management appraisal guidance.

Gee, K., & Burkhard, B. (2010). Cultural Ecosystem Services in the context of offshore windfarming: A case study from the west coast of Schleswig-Holstein. *Ecological Complexity, 7*(3), 349–358.

HM Government. (2011a, June). The Natural Choice: securing the value of nature. ISBN: 9780101808224.

HM Government. (2011b). The Marine Policy Statement. Reference: PB13654. ISBN: 978 0 10 851043 4. Available at: www.gov.uk/government/publications/uk-marine-policy-statement

HM Government. (2013). What nature can do for you. October 2010. Reference: PB13897. Available at: www.gov.uk/government/what-can-nature-do-for-you

HM Treasury. (2011). The Green Book. Appraisal and evaluation in central government 2011. London, UK: HM Treasury.

Jones, L., Angus, S., Cooper, A., Doody, P., Everard, M., Garbutt, A., Gilchrist, P., Hansom, J., Nicholls, R., Pye, K., Ravenscroft, N., Rees, S., Rhind, P., & Whitehouse, A. (2011). *Chapter 11: Coastal margin habitats*. UK National Ecosystem Assessment, UNEP-WCMC, Cambridge.

Kenter, J. O., Reed, M. S., Irvine, K. N., O'Brien, E., Brady, E., Bryce, R., Christie, M., Church, A., Cooper, N., Davies, A., Hockley, N., Fazey, I., Jobstvogt, N., Molloy, C., Orchard-Webb, J., Ravenscroft, N., Ryan, M., & Watson, V. (2014). UK National Ecosystem Assessment Follow-on. Work Package Report 6: Shared, plural and cultural values of ecosystems. UNEP-WCMC, LWEC, UK.

Liquete, C., Piroddi, C., Drakou, E., Gurney, L., Katsanevakis, S., Charef, A., & Egoh, B. (2013). Current status and future prospects for the assessment of marine and coastal ecosystem services: A systematic review. *PloS One, 8*(7), e67737.

MMO. (2014). Exploring the potential of using Official National Statistics (ONS) for marine planning. A report produced for the Marine Management Organisation, p. 80. MMO Project No: 1075. ISBN: 978-1-909452-37-4. Available at: https://www.gov.uk/government/publications/exploring-the-potential-of-using-office-for-national-statistics-ons-data-for-marine-planning-mmo-1075

MMO, & Marine Scotland. (2012a). A review of marine social and economic data: A report produced for the Marine Management Organisation and Marine Scotland, p. 45. MMO Project No: 1012. ISBN:978-1-909452-01-5. Available at: http://webarchive.nationalarchives.gov.uk/20140108121958/http:/www.marinemanagement.org.uk/evidence/1012b.htm

MMO, & Marine Scotland. (2012b). A critical review of tools and methods to apply marine social and economic data to decision-making. A report produced for the Marine Management Organisation and Marine Scotland, p. 58. MMO Project No: 1012. ISBN: 978-1-909452-02-2. Available at: http://www.scotland.gov.uk/Resource/0041/00412947.pdf

National Ecosystem Assessment (NEA). (2011). *The UK National Ecosystem Assessment technical report*. Cambridge: UNEP-WCMC.

Potts, T., O'Higgins, T., Mee, L., & Pita, C. (2011). Public perceptions of Europe's Seas – A policy brief. EU FP7 KNOWSEAS. Project. 2011, 23 pp., ISBN 0-9529089-3-X.

Potts, T., O'Higgins, T., & Hastings, E. (2012). Oceans of opportunity or rough seas? What does the future hold for developments in European marine policy? *Philosophical Transactions of the Royal Society A, 370*(1980), 5682–5700.

Potts, T., Burdon, D., Jackson, E., Atkins, J., Saunders, J., Hastings, E., & Langmead, O. (2014). Do marine protected areas deliver flows of ecosystem services to support human welfare? *Marine Policy, 44*, 139–148.

Saunders, J., Tinch, R., & Hull, S. (2010a). Valuing the marine estate and UK seas: An Ecosystem Services Framework [online]. The Crown Estate, 54 pp. ISBN: 978-1-906410-15-5. Available at: http://www.thecrownestate.co.uk/media/207041/valuing_marine_estate_uk_seas.pdf

Saunders, J., Tinch, R., Ozdemiroglu, E., & Hull, S. (2010b). Valuing the marine estate and UK seas: Initial Static Baseline Assessment [online]. The Crown Estate, 53 pp. ISBN: 978-1-906410-17-9. Available at: http://www.thecrownestate.co.uk/media/207037/initial_static_baseline_assessment_of_uk_seas_2010.pdf

Scott, A., Carter, C., Hölzinger, O., Everard, M., Rafaelli, D., Hardman, M., Baker, J., Glass, J., Leach, K., Wakeford, R., Reed, M., Grace, M., Sunderland, T., Waters, R., Corstanje, R., Glass, R., Grayson, N., Harris, J., & Taft, A. (2014). UK National Ecosystem Assessment Follow-on. Work Package Report 10: Tools – Applications, Benefits and Linkages for Ecosystem Science (TABLES). UNEP-WCMC, LWEC, UK.

Socha, Y. M. (2013). Out of cite, out of mind: The current state of practice, policy, and technology for the citation of data. CODATA-ICSTI Task Group on Data Citation Standards and Practices. *Data Science Journal, 12*, 76 pp.

UKMMAS (UK Marine Monitoring and Assessment Strategy Group). (2010). *Charting Progress 2: Productive seas evidence group feeder report*. Published by Defra on behalf of UKMMAS.

United Nations. (1987). Report of the World Commission on environment and development: Our common future. Annex to *document A/42/427* – Development and International Co-operation: Environment, New York.

# Part III
# Case Studies

# Chapter 9
# Linking Ecosystem Services of Marine Protected Areas to Benefits in Human Wellbeing?

**Justine Saunders, Tavis Potts, Emma Jackson, Daryl Burdon, Jonathan P. Atkins, Emily Hastings, Olivia Langmead, and Steve Fletcher**

## 9.1 Introduction

This chapter examines the potential relationship between ecosystem services provided by coastal ecosystems and the designation and management of Marine Protected Areas (MPAs). Ecosystem services are defined as the outputs of ecosystems from which people and society derive benefits (MEA 2005). The hypothesis is that significant relationships exist between the provision of a range of ecosystem services and the features protected by the designation of MPAs. Furthermore, this

J. Saunders (✉)
Fugro EMU Limited, 1 Mill Court, The Sawmills, Durley, Southampton SO32 2EJ, England
e-mail: jus_at_home@yahoo.co.uk

T. Potts
Department of Geography and Environment, The University of Aberdeen, Elphinstone Road, Aberdeen AB24 3UF, UK

E. Jackson
School of Medical and Applied Sciences, Central Queensland University, Bryan Jordan Drive, PO Box 1319, Gladstone, QLD 4680, Australia

D. Burdon
Institute of Estuarine and Coastal Studies, University of Hull, Hull HU6 7RX, UK

J.P. Atkins
Hull University Business School, University of Hull, Hull HU6 7RX, UK

E. Hastings
James Hutton Institute, Dundee, UK

O. Langmead
Marine & Coastal Policy Research Centre, Marine Institute, Plymouth University, Devon, UK

Marine Biological Association of the UK, Plymouth, UK

S. Fletcher
School of Marine Science and Engineering, Plymouth University, Devon, UK

 
R.K. Turner, M. Schaafsma (eds.), *Coastal Zones Ecosystem Services*, Studies in Ecological Economics 9, DOI 10.1007/978-3-319-17214-9_9

protection will maintain these features in good ecological condition and in some cases restore ecological function with positive effects on the delivery of ecosystem services and benefits to human wellbeing. While all coastal and marine habitats provide a range of service functions, the implementation of an MPA may result in improvements in the quality or supply of an ecosystem service where pressures upon protected features are well-managed and reduced.

Such benefits from MPA designation are dependent on the suite of management measures that are specified for the site and the extent of protection they offer as well as practical issues such as levels of enforcement of those measures. Within the European Union (EU), the establishment of a network of MPAs is required to meet obligations under a number of international agreements including the OSPAR Convention (in the NE Atlantic), the World Summit for Sustainable Development and the Convention on Biological Diversity. The OSPAR Commission's strategic objective with regard to biodiversity and ecosystems is to '*halt and prevent by 2020 further loss of biodiversity in the OSPAR maritime area, to protect and conserve ecosystems and to restore, where practicable, marine areas which have been adversely affected*'. The marine environment is currently subject to statutory protection by a number of European Marine Sites (EMS), established under European legislation, which include Special Areas of Conservation (SACs) under the Habitats and Species Directive (92/43/EEC) and Special Protection Areas (SPAs) under the Birds Directive (2009/147/EC). Together these sites contribute towards the Europe-wide *Natura* 2000 network and the OSPAR MPA network. The conservation objectives for SACs range from maintaining to recovering the Favourable Conservation Status of Annex I habitats and Annex II species. Wetlands of international importance may be protected under the Ramsar Convention while further national nature conservation exists in the form of Marine Nature Reserves (MNRs) and Sites or Areas of Special Scientific Interest (SSSIs and ASSIs).

Within the UK, the protection afforded by these sites will not ensure that the UK meets the conservation objectives set out in the UK Marine Policy Statement (March 2011). As such, the UK Government is committed to '*creating a UK-wide ecologically coherent network of MPAs as a key element of its wider work to recover and conserve the richness of our marine environment and wildlife*' by 2012. The establishment of further MPAs will also assist with the achievement of Good Environmental Status under the EU Marine Strategy Framework Directive (2008/56/EC). The fundamental purpose of MPAs in the UK is habitat and species conservation. However, in the context of MPAs providing benefits for society, accounting for the export of ecosystem services from protected sites is important (Roncin et al. 2008) and may be a key determinant of MPA success via community support. Identifying and valuing ecosystem services from MPAs can highlight the mix of services produced from marine systems in general and those that can potentially be enhanced by MPA management. This includes local-scale provisioning services (e.g. marine resources such as fisheries) to large and longer-scale regulatory processes that support human welfare (e.g. carbon sequestration). While conservation is the primary aim of MPAs, evidence on the benefits to human wellbeing can be an important factor in their social and political acceptability.

As the devolved administrations in the UK develop and implement policies to designate new MPAs, there is an opportunity to embed conservation objectives into policy related to the provision of ecosystem services. We argue that MPAs are able to influence ecosystem services and this is dependent on design concepts such as the scale of the site, the listed features, and management processes. Understanding the portfolio of services provided by features within MPAs will improve planning and management, particularly in the context of making site-specific or regional trade-offs over designation, or in understanding the benefits and impacts of setting conservation objectives and introducing measures to achieve them.

Specific questions explored in this chapter include:

- How are ecosystem services built into international network design concepts for MPAs?
- How are ecosystem service concepts included in new UK MPA policy?
- What ecosystem services may protected features (habitats and species) in the UK provide?
- How may site management affect the provision of ecosystem services from MPAs?

This chapter will focus on the United Kingdom (UK) and examine the contrasting approaches to MPA designation applied by the devolved administrations in England, Scotland, Wales and Northern Ireland. The approach and findings will be discussed within the wider European and international context.

## 9.2 Network Design and Ecosystem Services Delivery

In 2003, the OSPAR Commission issued guidance to EU Member States on developing an ecologically coherent network of MPAs, based upon a set of design principles. This guidance, while not binding, supports the national implementation of MPA programs focused on coherence and the connections between constituent parts of MPAs (OSPAR 2006). Ecological coherence supports the resilience of MPAs to pressures such as climate change. This guidance has been interpreted by several Member States in supporting the designation of MPA networks but direct reference to ecosystem services in designation processes is sporadic. Below we interpret the general provisions of network design based on the OSPAR design principles, and highlight where they touch upon ecosystem service linkages.

MPA network designs which deliver ecological coherence are in essence aiming to contribute to both the health and resilience of the wider environment, and to the delivery of many supporting ecosystem services. A functioning, ecologically-coherent network of MPAs "should interact with, and support, the wider environment as well as other MPAs although this is dependent on appropriate management to support good ecosystem health and function within and outside the MPAs"(OSPAR 2006).

### 9.2.1 Proportion

The aim is to preserve biodiversity through protecting a sufficient or 'adequate' (JNCC and Natural England 2010) proportion of habitats and species and thereby the inherent resilience of the system. This principle influences other design concepts such as representativity and replication and contributes to the overall resilience of the MPA network. There are a number of international policy statements on the minimum proportion of each feature that should be protected ranging from 10 % to 60 % (see OSPAR 2006 for a list) however the final proportion should take into account the quality and amount of the feature present and the management measures to be applied. Successful application of this design principle will enhance supporting services generated from a protected feature.

### 9.2.2 Representativity

To be representative, an MPA network needs to protect the range of marine biodiversity found in our seas. This can be achieved by grouping habitats and species into broad-scale habitat types and protecting examples of these across the MPA network. However, little research has been carried out to ascertain how well protecting specific areas of broad-scale habitats represents the range of marine biodiversity (but see Rondinini 2011). Under the OSPAR approach, the representativity principle also includes protecting those features of conservation importance (FOCI) that are known to be rare, threatened, or declining in our seas. While species diversity is protected, presumably other supporting services are also enhanced as a consequence (e.g. species interactions). In addition, cultural services are likely to be enhanced by protecting species known to be rare, threatened or declining (e.g. seahorses). Despite a wealth of literature identifying links between ecosystem function and biodiversity in terms of the diversity of functional groups (Loreau et al. 2001), there have been few attempts to represent functional groups.

### 9.2.3 Replication

Replication is the protection of the same feature across multiple sites within the MPA network, taking biogeographic variation into account. Replication across sites provides insurance against natural variability and spreads the risk of damaging events and long-term change affecting sites. Because it is considered that replication may boost resilience in species populations across their geographic ranges, it contributes to a range of supporting services underpinning the delivery of many other ecosystem services (Folke et al. 2004).

### 9.2.4 Size and Shape

The size and shape of an individual MPA influences the degree to which external factors in the wider environment may affect protected features and the loss of organisms across reserve boundaries (negative spill-over). Size and shape will contribute to the integrity of an MPA's individual features as well as the viability and resilience of the MPA overall. The size of a site should be sufficient to ensure protected features are self-sustaining throughout natural cycles of variation. This is most important to species with low mobility, since highly mobile species with extensive ranges will not benefit from protection if they are exploited or disturbed across their range. However, with mobile species, the conservation of important nursery and feeding areas can help to support their viability. Equally shape is considered important to reduce edge effects and designs should follow natural boundaries rather than cross them. If the viability of a feature is maintained then this principle will support all the services that the feature provides.

### 9.2.5 Connectivity

Connectivity is the extent to which populations in different parts of a species range are linked by the movement of eggs, larvae or other propagules, juveniles or adults (Palumbi 2003). This is the main principle that seeks to optimise spill-over effects of commercially targeted species from MPAs to the surrounding areas. The successful application of this principle is key to improving the provisioning services gained through MPAs, but in addition is likely to play a strong role in maintaining the functioning of surrounding areas in terms of supporting services. In addition, connectivity contributes to ecological coherence in the MPA network (OSPAR 2006) and is important for maintaining resilience in a system (Olds et al. 2012). However, due to the different dispersal traits of the multiple species protected by an MPA network, assessing minimum distances for the spacing of MPAs can be problematic (Roberts et al. 2010).

### 9.2.6 Management

Protection levels required within MPAs are determined by the nature conservation aspirations for the MPA features, as set out in the conservation objectives associated with a site. Levels of protection may range from highly protected areas where no extraction, deposition or other damaging activities are allowed, to areas where only minimal restrictions are needed to protect the features. Highly protected areas aim for a natural system to develop that could be used as a reference area to demonstrate the services that could be provided in a less exploited system. They also allow comparison with potentially impacted sites to assist in the management of marine activities. This aspect of the designation process is a key factor in realising improved flows of ecosystem services from MPAs, since a conservation objective of 'recover' will mean increased protection and could enhance ecosystem service provision, while one of 'maintain' is unlikely to lead to improvements in the condition of the feature, its functioning or service provision. It is recognised that some ecosystem services will only be improved under high levels of protection (Natural England and JNCC 2010).

## 9.3 To What Extent Are Ecosystem Services Incorporated into New MPA Policy in the UK?

All of the four government administrations in the UK are committed to the shared vision of 'clean, healthy, safe, productive and biologically diverse oceans and seas' expressed in their 2002 Safeguarding our Seas report and consider marine nature conservation to be an integral component of this vision. The Marine and Coastal Access Act 2009 (MCAA) builds on and improves protection of marine biodiversity by introducing a new type of MPA referred to as a Marine Conservation Zone (MCZ) for English and Welsh territorial waters and adjacent offshore UK waters. MCZs are designated to protect specific nationally important marine wildlife, habitats, geology and geomorphology. Similarly, the Marine (Scotland) Act 2010 provides for the designation of MPAs in Scottish territorial waters and adjacent UK Offshore waters and MCZs may be introduced in Northern Ireland territorial waters under the Marine (Northern Ireland) Act 2013. The various approaches adopted by the four administrations (as at 30 September 2014) are outlined further below.

### *9.3.1 England*

The vision for the English MPA network is "*to recover and protect the richness of our marine wildlife and environment*" (Natural England and JNCC 2010). The focus is not on protecting ecosystem services directly, but on biodiversity conservation. This is evident in the Features Of Conservation Interest (FOCI) which are dominated by rare, scarce or threatened species as opposed to those that are functionally important. However, some habitat FOCI have been selected for their importance in service provision, in particular their importance in the recruitment of fisheries (e.g. seagrass beds) or for supporting high biodiversity (e.g. maerl beds).

The approach taken to achieve representativity of species and habitats was to conserve broad-scale habitats (Jackson et al. 2008; JNCC and Natural England 2010). A numerical approach was applied that identified the percentages of broad-scale habitats that would encompass 70–80 % of the species found within those habitats (Rondinini 2011). Vulnerability assessments were carried out by the regional projects using feature sensitivity together with levels of activities to determine whether conservation objectives should be set to 'maintain' the feature in its current favourable condition or 'recover' to favourable condition. High levels of protection were also proposed through the identification of Reference Areas (RAs).

The selection of recommended MCZs (rMCZs) in all English seas and adjacent offshore territorial waters was undertaken through four regional projects. Ecosystem level information was provided, such as areas of high productivity, nursery areas, migration routes, feeding areas, spawning grounds and areas of high biodiversity. However, only one of the projects took this information into account in the selection of rMCZs and only when decisions were being made between sites of equivalent merit in terms of ecological and socio-economic interests. In practical terms, this meant that attributes which may arguably be most informative in terms of functioning and ecosystem service provision were not considered a priority.

Proposals for a network of 127 rMCZs were published by Defra in September 2011. The Impact Assessment that accompanied these sites addressed the costs and benefits of each site individually, ignoring the positive ecosystem level benefits associated with the network. Of these sites, 27 were designated on 21 November 2013, although not all species and habitats originally proposed for protection were taken through with each site. A further 95 sites have been held back, as well as all of the RAs, with future designation dependent on a review of their underlying evidence. Five sites were deemed unsuitable for designation based on the Impact Assessment and consultation process (Defra 2012): the socio-economic costs of four sites were considered to outweigh the ecological conservation advantages, and there was little additional conservation value in protecting the other site. On 27 February 2014, Defra announced that an additional 37 MCZs will be considered as potential candidates for consultation in 2015, with a final tranche of sites being designated by the end of 2016 to complete the contribution to the ecologically coherent network.

### *9.3.2 Scotland*

The designation process for MPAs in Scotland aims to 'protect marine biodiversity and ecosystems to ensure that natural environment, and the diversity of industries which depend upon it, is safeguarded for the future' (Scottish Government 2012). This statement is underpinned by the Marine (Scotland) Act 2010, where the duty of the Act is 'sustainable development and protection and enhancement of the health of the Scottish marine area' and that Ministers 'must act in the way best calculated to further the achievement of sustainable development, including the protection and, where appropriate, enhancement of the health of that area'. MPAs may be designated for three different purposes. Nature conservation MPAs should consider mitigation of climate change (a final regulatory ecosystem service). MPAs for demonstration and research purposes support the management of service flows as a policy objective, particularly in the provisioning context, but also in relation to cultural services such as education or research or illustrating the outcomes of sustainable practices. Historical MPAs are designed around 'historic assets' (e.g. the remains of a vessel). Ministers may 'have regard' for social and economic consequences of MPA designation.

Further support exists for the dual nature of MPAs to deliver conservation and ecosystem service functions. The Strategy for Marine Nature Conservation in Scotland's Seas (Marine Scotland 2011) identifies that industries and communities 'depend on a range of ecosystem services delivered by marine biodiversity' and that spatial protection can maximise the flow of benefits to society. The Strategy includes guidelines for identifying sites that provide a flow of services. Guideline 1c states that "this guideline should include consideration of features or locations providing ecosystem services which underpin key human activities/use of the marine environment." This means that sites may also be proposed for selection that do not necessarily contain key or threatened species but contribute ecological resources that provide a range of services and benefits.

On 24 July 2014, 17 MPAs were designated in Scottish territorial waters under the Marine (Scotland) Act and 13 MPAs were designated in adjacent offshore waters under the UK MCAA. An independent science review of the MPAs suggested that the network is coherent in terms of Scottish seas (Earnshaw et al. 2014). Currently management plans and conservation objectives are being discussed amongst stakeholders.

### 9.3.3 Wales

A range of marine habitats and species are currently protected by 125 MPAs that cover 36 % of Welsh seas. Therefore, the Welsh Government's approach to using the new MCZ power was to supplement the levels of protection within 3–4 existing MPAs rather than create new sites. The intention was for these sites to function ecologically, as naturally as possible in order to maximise the contribution they make to ecosystem recovery and resilience. It was argued the best way of achieving this was to afford sites a high level of protection; that is protection from extraction and deposition of living and non-living resources plus all other damaging or disturbing activities. The emphasis on biodiversity, functioning and resilience was more closely aligned with an ecosystem services approach than a focus on lists of features. High levels of protection within the Welsh MCZs may have enhanced provisioning services preferentially, and created productive areas where species enhance stocks in surrounding waters (see case studies). However, the 2012 consultation responses to the proposals expressed strong concerns they would create unacceptable socio-economic impacts and that there was little evidence of the benefits. These sites have since been withdrawn with recommendations to reassess the contribution of the existing network of MPAs to design principles such as resilience and connectivity. Should future MCZs be deemed necessary to supplement the network, they will be selected following a stakeholder-led process and the level of protection and site management will be determined on a site-by-site basis following a risk-based approach.

### *9.3.4 Northern Ireland*

In Northern Ireland, MCZs will be designated under Part 3 (Marine Protection) of the Marine (Northern Ireland) Act 2013 to protect rare, threatened or nationally important marine habitats, species and geological features and, along with existing marine sites, MCZs will assist in achieving an ecologically coherent network. The primary aim of the network is on nature conservation of sensitive and ecologically important species and habitats rather than specifically on ecosystem service provision. Although the Marine (Northern Ireland) Act 2013 does not highlight the protection of ecosystem services per se, the DoE's draft strategy for MPAs in Northern Ireland recognises that 'the marine environment has a significant value to society, through the goods and services it provides' (DoE 2013b).

The approach for site selection will be based on the OSPAR network design principles, with a focus on a list of Priority Marine Feature (PMF) habitats, limited/low mobility species, highly mobile species and geological/geomorphic features (DoE 2013a). It is of note that no broad-scale habitats have been proposed for protection within the Northern Ireland inshore region. Potential sites will be selected for formal designation following discussion and consultation with stakeholders. A five stage process is proposed: (1) Identify Area of Search (AoS); (2) Prioritise AoS based on quality of PMF contained; (3) Assess the size of the AoS to ensure this is sufficient to maintain the integrity of features protected; (4) Assess the effectiveness of managing features within the proposed AoS; and (5) Assess the ecological coherence to prioritise between different areas based on the contribution to the MPA network. Following public consultation of the draft guidance, a series of stakeholder workshops are planned for 2014/2015 with a final list of proposed MCZs being released for public consultation by 31 December 2015.

## 9.4 The Ecosystem Service Benefits of MPAs in the UK

The matrices below present an overview of the intermediate ecosystem services and final goods and benefits provided by different marine features of conservation importance, including both habitats (Table 9.1) and species (Table 9.2). The matrices follow the framework outlined in Chap. 2, which illustrates the flow of ecosystem services from components of the marine ecosystem (in this case each protected habitat and species) towards intermediate ecosystem services and the goods and benefits that humans derive from these services. In the matrices, separate ranking of final services from benefits was deemed unnecessary as contributions are inherently

**Table 9.1** Relative importance of designated habitats in providing intermediate ecosystem services and goods/benefits

| Feature Type[†] | EUNIS code | Feature | Intermediate services | | | | | | | | | | | Goods/Benefits | | | | | | | | | | | | | | |
|---|---|---|---|---|---|---|---|---|---|---|---|---|---|---|---|---|---|---|---|---|---|---|---|---|---|---|---|---|
| | Note: Eunis codes were identified using the JNCC EUNIS translation matrix. Some habitats do not have a direct relationship to the EUNIS code and this column should only be used as a guide. | (Bold type represents Broadscale habitats, normal type represents habitat FOCI) | Supporting services | | | | | | | Regulating services | | | | from Provisioning services | | | | | from Regulating services | | | | from Cultural services | | | | | |
| | | | Primary production | Larval and gamete supply | Nutrient cycling | Water cycling | Formation of species habitat | Formation of physical barriers | Formation of seascape | Biological control | Natural hazard regulation | Waste breakdown and detoxification | Carbon sequestration | Food (wild, farmed) | Fish feed (wild, farmed, bait) | Fertiliser and biofuels | Ornaments and aquaria | Medicines and blue biotechnology | Healthy climate | Prevention of coastal erosion | Sea defence | Waste burial / removal / neutralisation | Tourism and nature watching | Spiritual and cultural well-being | Aesthetic benefits | Education and Research | Physical health benefits | Psychological health benefits |
| **Existing Habitats protected under EU legislation** | | | | | | | | | | | | | | | | | | | | | | | | | | | | |
| E,EU,W | A1.1 | **High energy intertidal rock** | 3 | 2 | 3 | 1 | 2 | 4 | 1 | | 1 | 1 | 2 | 3 | 1 | 1 | | | 2 | 1 | 1 | 1 | 1 | 1 | 1 | 1 | 2 | 1 |
| E,EU,W | A1.2 | **Moderate energy intertidal rock** | 3 | 2 | 3 | 1 | 2 | 4 | | | 1 | 1 | 2 | 3 | 1 | 1 | | | 2 | 1 | 1 | 1 | 1 | 1 | 1 | 1 | 2 | 1 |
| E,EU,W | A1.3 | **Low energy intertidal rock** | 3 | 2 | 3 | 1 | 2 | 4 | | | 1 | 1 | 2 | 3 | 1 | 1 | | | 2 | 1 | 1 | 1 | 1 | 1 | 1 | 1 | 1 | 1 |
| E,W | A2.2 | **Intertidal sand and muddy sand** | 3 | 3 | 3 | 1 | 3 | 1 | 3 | | 3 | 1 | 2 | 1 | 2 | 1 | | | 2 | 3 | 3 | 1 | 1 | 1 | 3 | 1 | 3 | 3 |
| E,W | A2.3 | **Intertidal mud** | 3 | 3 | 3 | 1 | 1 | 1 | 1 | | 3 | 3 | 3 | 3 | 3 | 1 | | | 3 | 3 | 3 | 3 | 1 | 1 | 1 | 1 | 1 | 1 |
| E,EU | A2.4 | **Intertidal mixed sediments** | 3 | 3 | 3 | 1 | 3 | 1 | 1 | | 3 | 1 | 2 | 1 | 2 | 1 | | | 2 | 3 | 3 | 1 | 1 | 1 | 1 | 1 | | |
| E | A2.5 | **Coastal saltmarshes and saline reedbeds** | 2 | 3 | 3 | 1 | 3 | 3 | 3 | | 3 | 3 | 3 | 3 | 1 | 3 | | | 3 | 3 | 3 | 3 | 3 | 1 | 3 | 1 | 1 | 1 |
| EU,E,W | A2.6 | **Intertidal sediments dominated by aquatic angiosperms** | 2 | | 3 | 1 | 2 | 1 | 1 | | 1 | 3 | 1 | 3 | 1 | 3 | | | 1 | 1 | 1 | 1 | 1 | 1 | | 1 | | |
| EU,E,W | A2.7 | **Intertidal biogenic reefs** | 1 | 1 | 2 | 1 | 3 | 1 | 1 | | 2 | 1 | 1 | 2 | 2 | 2 | | | 1 | 2 | 2 | 2 | 1 | 1 | | 1 | | |
| EU,E,W | A3.1 | **High energy infralittoral rock** | 2 | 2 | | 1 | 2 | 4 | | | 1 | 1 | | 3 | 1 | 1 | | | | 1 | 1 | 1 | 1 | 1 | | 1 | 1 | |
| EU,E,W | A3.2 | **Moderate energy infralittoral rock** | 2 | 2 | | 1 | 2 | 4 | | | 1 | 1 | | 3 | 1 | 1 | | | | 1 | 1 | 1 | 1 | 1 | | 1 | 1 | |
| EU,E,W | A3.3 | **Low energy infralittoral rock** | 2 | 2 | | 1 | 2 | 4 | | | 1 | 1 | | | 1 | 1 | | | | 1 | 1 | 1 | 1 | 1 | | 1 | 1 | |
| EU,E,W | A4.1 | **High energy circalittoral rock** | 2 | 2 | | 1 | 2 | 1 | | | 1 | 1 | | 1 | 1 | 1 | | | | 1 | 1 | 1 | 1 | 1 | | 1 | 1 | |
| EU,E,W | A4.2 | **Moderate energy circalittoral rock** | 2 | 2 | | 1 | 2 | 1 | | | 1 | 1 | | 1 | 1 | 1 | | | | 1 | 1 | 1 | 1 | 1 | | 1 | 1 | |
| EU,E,W | A4.3 | **Low energy circalittoral rock** | 2 | 2 | | 1 | 2 | 1 | | | 1 | 1 | | 1 | 1 | 1 | | | | 1 | 1 | 1 | 1 | 1 | | 1 | 1 | |
| S | A5.1, A5.2 | **Offshore subtidal sands and gravels** | 1 | 1 | 1 | 1 | 1 | 1 | | 1 | 1 | 1 | 1 | 1 | 3 | 1 | 1 | 1 | 1 | 1 | 1 | 1 | 1 | 1 | | 1 | | |
| E,W | A5.2 | **Subtidal sand** | 3 | 3 | 3 | 1 | 3 | 1 | | | 3 | 1 | | 2 | 3 | 1 | | | | 3 | 3 | 1 | | 1 | | 1 | | |
| EU,E,W | A5.4 | **Subtidal mixed sediments** | 3 | 3 | 3 | 1 | 3 | 1 | | | 3 | 3 | | 2 | 3 | 1 | | | | 3 | 3 | 3 | | 1 | | 1 | | |
| EU,E,W | A5.5 | **Subtidal macrophyte-dominated sediment** | 3 | 2 | 2 | 1 | 2 | 1 | | | 1 | 2 | 2 | 3 | 1 | 3 | | | 2 | 1 | 1 | 2 | 1 | 1 | | 1 | | |
| EU,E,W | A5.6 | **Subtidal biogenic reefs** | 1 | 2 | 3 | 1 | 2 | 1 | | | 3 | 3 | 1 | 3 | 1 | 2 | 1 | | 1 | 2 | 2 | 1 | 1 | 1 | | 1 | | |
| EU | X02 | **Saline lagoons** | | 3 | 3 | 1 | 3 | | 1 | | | | | 1 | 1 | 1 | | | | | | | 1 | 1 | | 1 | 1 | |

| | | | | | | | | | | | | | | | | | | | | | | | | | | | | |
|---|---|---|---|---|---|---|---|---|---|---|---|---|---|---|---|---|---|---|---|---|---|---|---|---|---|---|---|---|
| EU, E, NI | A1.32 | Estuarine rocky habitats | 1 | 1 | | 1 | 2 | 1 | 1 | | 1 | 1 | | 1 | 1 | | | | | 1 | 1 | 1 | 1 | 1 | 1 | 1 | 1 | |
| EU | A1.44 | Submerged or partially submerged sea caves | | | | 1 | 1 | 1 | 1 | | 1 | 1 | 1 | | 1 | 1 | | | 1 | 1 | 1 | 1 | 1 | 1 | | 1 | | |
| E,W, NI | A4.12 | Fragile sponge&anthozoan communities on subtidal rocky habitats | | 1 | | 1 | 3 | 1 | | 1 | 1 | 1 | | 3 | 1 | 1 | | | | 1 | 1 | 1 | 3 | 1 | | 1 | 1 | |
| W | A4.131, A4.2122 | Subtidal rock with Ross 'coral' *Pentapora foliacea* | | 1 | | 1 | 2 | 1 | | 1 | | 1 | | 1 | 1 | 1 | | | | | | 1 | 3 | 1 | | 1 | 1 | |
| E | A4.22 | Ross worm *Sabellaria spinulosa* reefs | | 1 | 1 | 1 | 3 | 1 | | 1 | 1 | 1 | 1 | 1 | 1 | | | | 1 | 1 | 1 | 1 | 1 | 1 | | 1 | | |
| All | A5.51 | Maerl beds | 3 | 1 | 1 | 1 | 3 | 1 | 1 | 1 | | | | 3 | 1 | 1 | 1 | | | | | | 1 | 1 | 1 | 1 | | 1 |
| S | A5.5112 | Maerl or coarse shell gravel with burrowing sea cucumbers | 3 | 1 | 1 | 1 | 3 | 1 | 1 | 1 | | | | 3 | 1 | 1 | 1 | | | | | | 1 | 1 | | 1 | | |
| All | A5.53, A5.545, A2.61 | Seagrass beds | 2 | 1 | 2 | 1 | 3 | 1 | 1 | 2 | 1 | 2 | 2 | 3 | 1 | 2 | 2 | 1 | 2 | 2 | 2 | 2 | 2 | 1 | 2 | 1 | | |
| EU | A5.71 | Submarine structures made by leaking gases | | | 1 | 1 | 3 | 1 | | | 1 | 1 | 1 | 1 | 1 | 1 | | 1 | 1 | 1 | 1 | 1 | | 1 | | 1 | | |
| **New habitats proposed under new MPA legislation** | | | | | | | | | | | | | | | | | | | | | | | | | | | | |
| E,W | A2.1 | **Intertidal coarse sediment** | 1 | 3 | 1 | 1 | 3 | 1 | 1 | | 3 | 1 | | 1 | 2 | 1 | 1 | | | 3 | 3 | 1 | 1 | 1 | 1 | 1 | 1 | |
| E,W | A5.1 | **Subtidal coarse sediment** | 3 | 3 | 3 | 1 | 3 | 1 | | | 3 | 1 | | 2 | 3 | 1 | | | | 3 | 3 | 1 | | 1 | | 1 | | |
| E,W | A5.3 | **Subtidal mud** | 3 | 3 | 3 | 1 | 3 | 1 | | | 3 | 3 | | 2 | 3 | 1 | | | | 3 | 3 | 3 | | 1 | | 1 | | |
| W | A5.4, A5.3 | **Subtidal mixed muddy sediments** | 3 | 3 | 3 | 1 | 3 | 1 | | | 3 | 3 | | 2 | 3 | 1 | | | | 3 | 3 | 3 | | 1 | | 1 | | |
| S | A7.4, A7.7 | **Salinity fronts** | 1 | 1 | 1 | 1 | 1 | 1 | | 1 | 1 | 2 | 1 | 1 | 1 | 1 | 1 | 1 | 1 | 1 | 1 | 2 | 1 | 1 | | 1 | | |
| S | Various | **Low or variable salinity habitats** | 1 | 1 | 1 | 1 | 1 | 1 | 1 | 1 | 1 | 1 | 1 | 1 | 1 | 1 | 1 | 1 | 1 | 1 | 1 | 1 | 1 | 1 | 1 | 3 | 2 | |
| E,W | A1.2142, A3.2112 | Intertidal under boulder communities | 1 | 1 | | 1 | 2 | 1 | | | 1 | 1 | | 2 | 1 | 1 | | | | 1 | 1 | 1 | 1 | 1 | 1 | 1 | | |
| E | A1.127, A1.223, A4.231 | Peat and clay exposures | | 1 | | 1 | 2 | 1 | | | | | | | 1 | | | | | | | | 1 | 1 | | 1 | | |
| S | A1.325 | Sea loch egg wrack beds | 1 | 1 | 1 | 1 | 1 | 1 | 1 | 1 | 1 | 1 | 1 | 1 | 1 | 1 | 1 | 1 | 1 | 1 | 1 | 1 | 1 | 1 | | 1 | | |
| E | A1.441, B3.114, B3.115 | Littoral chalk communities | 1 | 1 | | 1 | 3 | 1 | | | | 1 | | | 1 | 1 | | | | | | 1 | 1 | 1 | | 1 | | |
| E,S,W, NI | A2.2, A2.7, A5.6 | Blue Mussel beds | 1 | 1 | 1 | 1 | 1 | 1 | 1 | 1 | 1 | 3 | 1 | 2 | 1 | 2 | 1 | 1 | 1 | 1 | 1 | 1 | 1 | 1 | | 1 | | |
| NI | A2.23 or A5.2 | Stable sands with associated fauna | 1 | 1 | 1 | 1 | 2 | 2 | 2 | | 3 | 1 | 1 | 1 | 2 | 1 | | | 1 | 1 | 1 | 1 | 1 | 1 | 1 | 1 | 1 | |
| E,W | A2.71 | Honeycomb worm *Sabellaria alveolata* reef | | 1 | 1 | 1 | 3 | 1 | | 1 | 2 | 1 | 1 | 1 | 2 | | | | 1 | 1 | 1 | 1 | 1 | 1 | | 1 | | |
| S | A3.126, A3.213 | Tide-swept algal communities (Laminaria hyperborea, Halidrys siliquosa) | 1 | 1 | 1 | 1 | 1 | 1 | 1 | 1 | 1 | 1 | 1 | 1 | 1 | 1 | | 1 | 1 | 1 | 1 | 1 | 1 | 1 | 1 | 1 | | |
| S | A3.126, A3.213, A1.15 | Tide-swept algal communities | 1 | 1 | 1 | 1 | 1 | 1 | 1 | 1 | 1 | 1 | 1 | 1 | 1 | 1 | | 1 | 1 | 1 | 1 | 1 | 1 | 1 | 1 | 1 | | |
| S | A4.133, A4.211 | Northern sea fan and sponge communities | | | | 1 | 1 | 1 | 1 | | 1 | 1 | | | 1 | 1 | | | | 1 | 1 | 1 | | | 1 | | 1 | |
| E | A4.23 | Subtidal chalk | | 1 | | 1 | 2 | 1 | | | | 1 | | | 1 | 1 | | | | | | 1 | 1 | 1 | 1 | 1 | | |
| E | A5.12, A5.13 | Subtidal sands and gravels | 1 | 1 | 1 | 1 | 2 | 1 | | | | 1 | | 2 | 3 | 1 | | | | | | 1 | 1 | 1 | | 1 | | |
| S | A5.133 | Shallow tide-swept coarse sands with burrowing bivalves (Morella sp.) | | | 1 | 1 | 1 | 1 | | | | 1 | | 1 | 1 | 1 | | | | | | 1 | | 1 | | 1 | | |
| E,S, NI | A5.361 | Sea-pen and burrowing megafauna communities | 1 | 1 | 2 | 1 | 3 | 1 | 1 | 1 | 1 | 1 | 1 | 2 | 1 | 1 | 1 | 1 | 1 | 1 | 1 | 1 | 1 | 1 | 1 | 1 | | |
| S | A5.371 | Inshore deep mud with burrowing heart urchins | 3 | 3 | 3 | 1 | 3 | 1 | | | 3 | 1 | | 2 | 1 | 1 | | | | 3 | 3 | 3 | | 1 | | 1 | | |
| W, NI | A6.5 | Mud habitats in deep water | 3 | 3 | 3 | 1 | 3 | 1 | | | 3 | 1 | | 2 | 3 | 1 | | | | 3 | 3 | 3 | | 1 | | 1 | | |
| E,W | A5.43, A2.41, A2.42 | Sheltered muddy gravels | | 1 | 1 | 1 | 2 | 1 | | | | 1 | 1 | 2 | 1 | 1 | | | 1 | | | 1 | 1 | 1 | | 1 | | |
| E,S | A5.434 | Flame/ File shell beds | 1 | 1 | 3 | 1 | 3 | 1 | 1 | 1 | 3 | 1 | 1 | 3 | 1 | 1 | | | 1 | 1 | 1 | 1 | 1 | 1 | 1 | 1 | | |
| E,S,W, NI | A5.435 | Native Oyster *Ostrea edulis* beds | | 1 | 1 | 1 | 1 | 1 | 1 | 1 | 1 | 1 | 1 | 3 | 1 | 1 | | | 1 | 1 | 1 | 1 | 1 | 1 | 1 | 1 | | |
| S | A5.52 | Kelp and seaweed communities on sublittoral sediment | 1 | 1 | 1 | 1 | 1 | 1 | 1 | 1 | 1 | 1 | 1 | 1 | 1 | 1 | 1 | 1 | 1 | 1 | 1 | 1 | 1 | 1 | 1 | 1 | 1 | |
| E,S,W, NI | A5.62 | Horse mussel (Modiolus modiolus) beds | 1 | 1 | 1 | 1 | 3 | 1 | 1 | 1 | 1 | 1 | 1 | 1 | 1 | 1 | 1 | 3 | 1 | 1 | 1 | 1 | 1 | 1 | 1 | 1 | | |

(continued)

Table 9.1 (continued)

| Feature Type[†] | EUNIS code | Feature | Intermediate services | | | | | | | | | | | Goods/Benefits | | | | | | | | | | | | | | |
|---|---|---|---|---|---|---|---|---|---|---|---|---|---|---|---|---|---|---|---|---|---|---|---|---|---|---|---|---|
| | Note: Eunis codes were identified using the JNCC EUNIS translation matrix. Some habitats do not have a direct relationship to the EUNIS code and this column should only be used as a guide. | (Bold type represents Broadscale habitats, normal type represents habitat FOCI) | Supporting services | | | | | | | Regulating services | | | | from Provisioning services | | | | | from Regulating services | | | | from Cultural services | | | | | |
| | | | Primary production | Larval and gamete supply | Nutrient cycling | Water cycling | Formation of species habitat | Formation of physical barriers | Formation of seascape | Biological control | Natural hazard regulation | Waste breakdown and detoxification | Carbon sequestration | Food (wild, farmed) | Fish feed (wild, farmed, bait) | Fertiliser and biofuels | Ornaments and aquaria | Medicines and blue biotechnology | Healthy climate | Prevention of coastal erosion | Sea defence | Waste burial / removal / neutralisation | Tourism and nature watching | Spiritual and cultural well-being | Aesthetic benefits | Education and Research | Physical health benefits | Psychological health benefits |
| E, NI | A5.63 | Cold-water coral reefs | | 1 | 1 | 1 | 2 | 3 | | | | | 1 | 1 | 1 | 1 | | | 1 | | | | 1 | 1 | 2 | 1 | | |
| E,S | A6.61 | Coral Gardens | 1 | 1 | 1 | 1 | 3 | 1 | 1 | 1 | 1 | 1 | 1 | 1 | 1 | 1 | 3 | 1 | 1 | 1 | 1 | 1 | 1 | 1 | 1 | 1 | | |
| S | A6.75 | Carbonate mound communities | 1 | 1 | 1 | 1 | 1 | 1 | | 1 | 1 | 1 | 1 | 1 | 1 | 1 | 1 | 1 | 1 | 1 | 1 | 1 | 1 | 1 | | 1 | | |
| E,W | Various | Tide-swept channels | | 1 | | 1 | 1 | 1 | | | 1 | | | | 1 | 1 | | | | 1 | 1 | | 1 | 1 | | 1 | | |
| W | Various | Sediment habitats with long lived bivalves | | | 2 | 1 | 1 | 1 | | 1 | | 1 | | 1 | 1 | 1 | | | | | | 1 | 1 | 1 | | 1 | | |
| E | N/A | Areas of high planktonic primary productivity | 3 | 1 | 3 | 1 | 1 | 1 | | 1 | 1 | | 2 | 2 | 3 | 1 | | | 2 | 1 | 1 | 1 | 1 | 1 | | 1 | | |

**Scale of ecosystem service supplied relative to other features**

- # Significant contribution
- # Moderate contribution
- # Low contribution
- # No or negligible ESP
- Not assessed

**Confidence in evidence**

- 4 Obvious non-linkages
- 3 UK-related, peer-reviewed literature
- 2 Grey or overseas literature
- 1 Expert opinion or Obvious
- Not assessed

**Feature type[†]**

- S Scottish MPA search feature
- E English MCZ feature
- W Welsh HP MCZ feature
- NI Northern Ireland MCZ feature
- EU EU Habitats Directive Annex 1 feature or sub-feature

**Table 9.2** Relative importance of designated species in providing intermediate ecosystem services and goods/benefits

| Feature Type † | Species Names | Scientific Name | Intermediate Services | | | | | | | | | | | Goods/Benefits | | | | | | | | | | | | | | |
|---|---|---|---|---|---|---|---|---|---|---|---|---|---|---|---|---|---|---|---|---|---|---|---|---|---|---|---|---|
| | | | Supporting services | | | | | | | Regulating services | | | | from Provisioning services | | | | | from Regulating | | | | from Cultural services | | | | | |
| | | | Primary production | Larval and gamete supply | Nutrient cycling | Water cycling | Formation of species habitat | Formation of physical barriers | Formation of seascape | Biological control | Natural hazard regulation | Waste breakdown and detoxification | Carbon sequestration | Food (wild, farmed) | Fish feed (wild, farmed, bait) | Fertiliser and biofuels | Ornaments and aquaria | Medicines and blue biotechnology | Healthy climate | Prevention of coastal erosion | Sea defence | Waste burial / removal / neutralisation | Tourism and nature watching | Spiritual and cultural well-being | Aesthetic benefits | Education | Physical health benefits | Psychological health benefits |
| **Existing species protected under EU legislation** | | | | | | | | | | | | | | | | | | | | | | | | | | | | |
| EU | Allis shad | *Alosa alosa* | 1 | 3 | 3 | 1 | | 1 | | | 1 | 1 | 1 | 1 | 1 | 1 | | | 1 | 1 | 1 | 1 | 1 | | | 1 | | 1 |
| EU | Twaite shad | *Alosa fallax* | 1 | 3 | 3 | 1 | | 1 | | | 1 | 1 | 1 | 1 | 1 | 1 | | | 1 | 1 | 1 | 1 | 1 | | | 1 | | 1 |
| EU | Atlantic salmon | *Salmo salar* | 1 | 3 | 3 | 1 | | 1 | | | 1 | 1 | 1 | 3 | 1 | 1 | | | 1 | 1 | 1 | 1 | 3 | 1 | 2 | 1 | 2 | 1 |
| EU | Sea lamprey | *Petromyzon marinus* | 1 | 3 | | 1 | | 1 | 1 | 2 | 1 | 1 | 1 | 2 | 1 | 1 | | | 1 | 1 | 1 | 1 | 1 | | | 1 | 1 | 1 |
| EU | River lamprey | *Lampetra fluviatilis* | 1 | 3 | | 1 | | 1 | | 2 | 1 | 1 | 1 | 2 | 1 | 1 | | | 1 | 1 | 1 | 1 | 1 | | | 1 | 1 | 1 |
| EU | Grey seal | *Halichoerus grypus* | 1 | | | 1 | 1 | 1 | 1 | | 1 | 1 | 1 | 1 | 1 | 1 | 1 | | 1 | 1 | 1 | 1 | 3 | | 2 | 3 | 1 | 1 |
| EU | Common seal | *Phoca vitulina* | 1 | | | 1 | 1 | 1 | 1 | | 1 | 1 | 1 | 1 | 1 | 1 | 1 | | 1 | 1 | 1 | 1 | 3 | | 2 | 3 | 1 | 3 |
| EU, S | Bottlenose dolphin | *Tursiops truncatu* | 1 | 1 | 1 | 1 | 1 | 1 | 1 | 1 | 1 | 1 | 1 | 1 | 1 | 1 | 1 | 1 | 1 | 1 | 1 | 1 | 3 | 3 | 3 | 1 | 2 | 1 |
| EU, S, NI | Harbour porpoise | *Phocoena phocoena* | 1 | 1 | 1 | 1 | 1 | 1 | 1 | 1 | 1 | 1 | 1 | 1 | 1 | 1 | 1 | 1 | 1 | 1 | 1 | 1 | 3 | 3 | 3 | 1 | 1 | 1 |
| EU | Otter | *Lutra lutra* | 1 | | | 1 | | 1 | 1 | | | 1 | 1 | 1 | 1 | 1 | | | 1 | | 1 | 1 | 3 | 3 | 1 | 3 | 1 | 1 |
| **New species proposed for protection under new MPAs - highly mobile** | | | | | | | | | | | | | | | | | | | | | | | | | | | | |
| E | Smelt | *Osmerus eperlanus* | 1 | 3 | | 1 | | 1 | | | 1 | 1 | 1 | 3 | | 1 | 3 | | 1 | 1 | 1 | 1 | 3 | 1 | | 1 | 1 | 1 |
| E | European eel | *Anguilla anguilla* | 1 | 3 | 1 | 1 | | 1 | | | 1 | 1 | 1 | 3 | | 1 | 3 | | 1 | 1 | 1 | 1 | 1 | 1 | 1 | 1 | 1 | 1 |
| S | Blue ling | *Molva dypterygia* | 1 | 1 | 1 | 1 | 1 | 1 | | | 1 | 1 | 1 | 1 | | 1 | 1 | 1 | 1 | 1 | 1 | 1 | 1 | 1 | 1 | 1 | 1 | 1 |
| S | Orange roughy | *Hoplostethus atlanticus* | 1 | | | 1 | 1 | 1 | | | 1 | 1 | 1 | 2 | | 1 | 1 | | 1 | 1 | 1 | 1 | 1 | 1 | | 1 | 1 | 1 |
| S | Sandeels | | 1 | | | 1 | | | | | 1 | 1 | 1 | 1 | 3 | 1 | | | 1 | 1 | 1 | 1 | 1 | 1 | 1 | 1 | 1 | 1 |
| NI | Sole | *Solea solea* | 1 | 3 | | 1 | | 1 | | | 1 | 1 | 1 | 3 | | 1 | | | 1 | 1 | 1 | 1 | 1 | 1 | | | 1 | 1 |
| NI | Plaice | *Pleuronectes platessa* | 1 | 3 | | 1 | | 1 | | | 1 | 1 | 1 | 3 | | 1 | | | 1 | 1 | 1 | 1 | 1 | 1 | | | 1 | 1 |
| E | Undulate ray | *Raja undulata* | 1 | | | 1 | | 1 | | | 1 | 1 | 1 | 1 | | 1 | | | 1 | 1 | 1 | 1 | | | | 1 | 1 | 1 |
| NI | Spotted ray | *Raja montagui* | 1 | 3 | | 1 | | 1 | | | 1 | 1 | 1 | 2 | | 1 | | | 1 | 1 | 1 | 1 | 1 | 1 | | | 1 | 1 |
| NI | Thornback ray | *Raja clavata* | 1 | 3 | | 1 | | 1 | | | 1 | 1 | 1 | 2 | | 1 | | | 1 | 1 | 1 | 1 | 1 | 1 | | | 1 | 1 |
| NI | Cuckoo ray | *Leucoraja naevus* | 1 | 3 | | 1 | | 1 | | | 1 | 1 | 1 | 2 | | 1 | | | 1 | 1 | 1 | 1 | 1 | 1 | | | 1 | 1 |
| S, NI | Common skate | *Dipturus batis* | 1 | 1 | 1 | 1 | 1 | 1 | 1 | | 1 | 1 | 1 | 3 | | 1 | | | 1 | 1 | 1 | 1 | 1 | 1 | 1 | 1 | 2 | 1 |
| S | Basking shark | *Cetorhinus maximus* | 1 | 1 | 1 | 1 | 1 | 1 | 1 | | 1 | 1 | 1 | 1 | | 1 | 1 | 1 | 1 | 1 | 1 | 1 | 1 | 1 | 2 | 1 | 1 | 1 |
| S | Minke whale | *Balaenoptera acutorostrata* | 1 | 1 | 1 | 1 | 1 | 1 | 1 | 1 | 1 | 1 | 1 | 1 | | 1 | 1 | 1 | 1 | 1 | 1 | 1 | 2 | 1 | 1 | 1 | 1 | 1 |
| S | Risso's dolphin | *Grampus griseus* | 1 | 1 | 1 | 1 | 1 | 1 | 1 | 1 | 1 | 1 | 1 | 1 | | 1 | 1 | 1 | 1 | 1 | 1 | 1 | 1 | 1 | 1 | 1 | 1 | 1 |
| S | White-beaked dolphin | *Lagenorhynchus albirostris* | 1 | 1 | 1 | 1 | 1 | 1 | 1 | 1 | 1 | 1 | 1 | 1 | | 1 | 1 | 1 | 1 | 1 | 1 | 1 | 1 | 1 | 1 | 1 | 1 | 1 |
| **New species proposed for protection under new MPAs - Low or limited mobility species** | | | | | | | | | | | | | | | | | | | | | | | | | | | | |
| E, W | Peacock's tail | *Padina pavonica* | 2 | | 1 | 1 | 2 | 1 | | | 1 | | | 1 | | 1 | | 1 | | 1 | 1 | | | | | 1 | | 1 |
| E, W | Burgundy maerl paint | *Cruoria cruoriaeformis* | 1 | 1 | 1 | 1 | 1 | 1 | | 1 | 1 | | | 1 | | 1 | 1 | | | 1 | 1 | | 1 | 1 | | 1 | | 1 |
| E, W | Grateloup's little-lobed weed | *Grateloupia montagnei* | 1 | 1 | 1 | 1 | 1 | 1 | | 1 | 1 | | | 1 | | 1 | 1 | | | 1 | 1 | | 1 | 1 | | 1 | | 1 |
| E, W | Coral maerl | *Lithothamnion corallioides* | 3 | 1 | 1 | 1 | 3 | 1 | 1 | | | | | 3 | | 1 | 1 | | | | | | | 1 | | 1 | | 1 |
| E, W | Common maerl | *Phymatolithon calcareum* | 3 | 1 | 1 | 1 | 3 | 1 | 1 | | | | | 3 | | 1 | 1 | | | | | | | 1 | | 1 | | 1 |
| W | Bearded red seaweed | *Anotrichium barbatum* | 1 | | | 1 | | 1 | | | 1 | | | 1 | | 1 | | | | 1 | 1 | | | 1 | | 1 | | 1 |
| NI | A brown alga | *Ascophyllum nodosum ecad mackayii* | 3 | | 1 | | 2 | 1 | | | 1 | | | | | 1 | | | | 1 | 1 | | | | | | | 1 |
| NI | A red alga | *Atractophora hypnoides* | 3 | | 1 | | 2 | 1 | | | 1 | | | | | 1 | | | | 1 | 1 | | | | | | | 1 |
| E | Tentacled lagoon-worm | *Alkmaria romijni* | 1 | | | 1 | | 1 | | | 1 | | 1 | 1 | | 1 | | | 1 | 1 | 1 | | | 1 | | 1 | | 1 |
| E | Lagoon sandworm | *Armandia cirrhosa* | 1 | | | 1 | | 1 | | | 1 | | 1 | 1 | | 1 | | | 1 | 1 | 1 | | | 1 | | 1 | | 1 |

(continued)

**Table 9.2** (continued)

| Feature Type † | Species Names | Scientific Name | Intermediate Services: Supporting services: Primary production | Larval and gamete supply | Nutrient cycling | Water cycling | Formation of species habitat | Formation of physical barriers | Formation of seascape | Intermediate Services: Regulating services: Biological control | Natural hazard regulation | Waste breakdown and detoxification | Carbon sequestration | Goods/Benefits: from Provisioning services: Food (wild, farmed) | Fish feed (wild, farmed, bait) | Fertiliser and biofuels | Ornaments and aquaria | Medicines and blue biotechnology | Goods/Benefits: from Regulating: Healthy climate | Prevention of coastal erosion | Sea defence | Waste burial / removal / neutralisation | Goods/Benefits: from Cultural services: Tourism and nature watching | Spiritual and cultural well-being | Aesthetic benefits | Education | Physical health benefits | Psychological health benefits |
|---|---|---|---|---|---|---|---|---|---|---|---|---|---|---|---|---|---|---|---|---|---|---|---|---|---|---|---|---|
| E | Giant goby | *Gobius cobitis* | 1 | | | 1 | | 1 | | | 1 | | 1 | 2 | | 1 | | | 1 | 1 | 1 | | 1 | | | 1 | | 1 |
| E | Couch's goby | *Gobius couchi* | 1 | | | 1 | | 1 | | | 1 | | 1 | 3 | | 1 | | | 1 | 1 | 1 | | 1 | | | 1 | | 1 |
| E | Long snouted seahorse | *Hippocampus guttulatus* | 1 | 3 | | 1 | | 1 | 1 | | 1 | | 1 | 2 | | 1 | 1 | 1 | 1 | 1 | 1 | | 1 | 3 | | 3 | | 3 |
| E | Short snouted seahorse | *Hippocampus hippocampus* | 1 | 3 | | 1 | | 1 | 1 | | 1 | | 1 | 2 | | 1 | 1 | 1 | 1 | 1 | 1 | | 1 | 3 | | 3 | | 3 |
| E | Trembling sea mat | *Victorella pavida* | 1 | | | 1 | | 1 | | | 1 | | | 1 | | 1 | | | | 1 | 1 | | | 1 | | 1 | | 1 |
| NI | Ross coral / Potato crisp bryozoan | *Pentapora foliacea* | 1 | | | 1 | 2 | 1 | 2 | | 1 | | | | | 1 | | | | 1 | 1 | | 1 | | | | | 1 |
| NI | Football sea squirt | *Diazona violacea* | 1 | | | 1 | 2 | 1 | 2 | | 1 | | 1 | | | 1 | | | 1 | 1 | 1 | | 1 | | | | | 1 |
| S, W | Burrowing sea anemone aggregations | *Arachnanthus sarsi* | 1 | | 1 | 1 | 1 | 1 | | | 1 | | 1 | 1 | | 1 | | | 1 | 1 | 1 | | 1 | 1 | | 1 | | 1 |
| E | Sea-fan anemone | *Amphianthus dohrnii* | 1 | | | 1 | 1 | 1 | 1 | | 1 | | 1 | 1 | | 1 | | | 1 | 1 | 1 | | | 1 | | 1 | | 1 |
| E,W | Pink sea-fan | *Eunicella verrucosa* | 1 | | | 1 | 1 | 1 | 1 | | 1 | | 1 | 1 | | 1 | | | 1 | 1 | 1 | | 1 | 1 | | 3 | | 1 |
| E,W | Kaleidoscope jellyfish | *Haliclystus auricula* | 1 | | | 1 | | 1 | | | 1 | | 1 | 1 | | 1 | | | 1 | 1 | 1 | | | | | 1 | | 1 |
| E | Sunset cup coral | *Leptopsammia pruvoti* | 1 | | | 1 | 2 | 1 | 1 | | 1 | | | 1 | | 1 | | | | 1 | 1 | | 3 | | | 1 | | 1 |
| E,W | Stalked jellyfish | *Lucernariopsis campanulata* | 1 | | | 1 | | 1 | | | 1 | | | 1 | | 1 | | | | 1 | 1 | | | | | 1 | | 1 |
| E | St. John's jellyfish | *Lucernariopsis cruxmelitensis* | 1 | | | 1 | | 1 | | | 1 | | | 1 | | 1 | | | | 1 | 1 | | | | | 1 | | 1 |
| E | Starlet sea anemone | *Nematostella vectensis* | 1 | | | 1 | | 1 | | | 1 | | | 1 | | 1 | | | | 1 | 1 | | | | | 2 | | 1 |
| E | Lagoon sand shrimp | *Gammarus insensibilis* | 1 | | | 1 | | 1 | | | 1 | | 1 | 1 | | 1 | | | 1 | 1 | 1 | | | | | 1 | | 1 |
| E | Gooseneck barnacle | *Pollicipes pollicipes* | 1 | | | 1 | 3 | 1 | 1 | | 1 | | 1 | 3 | | 1 | | | 1 | 1 | 1 | | 1 | | | 1 | | 1 |
| E, S | Spiny lobster | *Palinurus elephas* | 1 | 1 | 1 | 1 | 1 | 1 | 1 | | 1 | 1 | 1 | 3 | | 1 | 1 | 1 | 1 | 1 | 1 | 1 | 1 | 1 | 1 | 1 | 1 | 1 |
| E, S, W, NI | Ocean quahog | *Arctica islandica* | 1 | | | 1 | | 1 | 1 | | 1 | | | 3 | | 1 | | | | 1 | 1 | | 1 | 1 | | 1 | | 1 |
| E, S, W | Fan mussel | *Atrina pectinata* | 1 | | 1 | 1 | 1 | 1 | 1 | | 1 | 1 | | 1 | | 1 | 1 | | | 1 | 1 | 1 | 1 | 1 | | 1 | | 1 |
| E, W | Native oyster | *Ostrea edulis* | 1 | | 1 | 1 | 1 | 1 | 1 | 1 | 1 | 1 | 1 | 3 | | 1 | | | 1 | 1 | 1 | 1 | 1 | 1 | | 1 | 1 | 1 |
| W | Smooth venus clam | *Callista chione* | 1 | | | 1 | | 1 | | | 1 | | | 1 | | 1 | | 2 | | 1 | 1 | | | 1 | | 2 | | 1 |
| S | Heart cockle aggregations | *Glossus humanus* | 1 | | | 1 | | 1 | | | 1 | | | 1 | | 1 | | | | 1 | 1 | | 1 | 1 | | 1 | | 1 |
| NI | Queen scallop | *Aequipecten opercularis* | 1 | | 1 | 1 | | 1 | | | 1 | | | 3 | | 1 | | | | 1 | 1 | | | | | | 1 | 1 |
| NI | Brackish cockle | *Cerastoderma glaucum* | 1 | | 1 | 1 | | 1 | | | 1 | | | 3 | | 1 | | | | 1 | 1 | | | | | | | 1 |
| E | Defolin's lagoon snail | *Caecum armoricum* | 1 | | | 1 | | 1 | | | 1 | | 1 | 1 | | 1 | | | 1 | 1 | 1 | | | 1 | | 1 | | 1 |
| E | Sea snail | *Paludinella littorina* | 1 | | | 1 | | 1 | | | 1 | | 1 | 1 | | 1 | | | 1 | 1 | 1 | | | | | 1 | | 1 |
| E | Lagoon sea slug | *Tenellia adspersa* | 1 | | | 1 | | 1 | 1 | | 1 | | 1 | 1 | | 1 | | | 1 | 1 | 1 | | | | | 1 | | 1 |
| S | Northern feather star aggregations on mixed substrata | *Leptometra celtica* | 1 | | | 1 | | 1 | 1 | | 1 | | | 1 | | 1 | | | | 1 | 1 | | 1 | 1 | | 1 | | 1 |
| NI | Cushion star | *Porania pulvillus* | 1 | | | 1 | | 1 | | | 1 | | 1 | | | 1 | | | 1 | 1 | 1 | | | | | | | 1 |
| NI | Drummond's sea cucumber | *Thyonidium drummondi* | 1 | | | 1 | | 1 | | | 1 | | 1 | | | 1 | | | 1 | 1 | 1 | | | | | | | 1 |

**Scale of ecosystem service supplied reative to other features**

| | |
|---|---|
| # (black) | Significant contribution |
| # (dark grey) | Moderate contribution |
| # (light grey) | Low contribution |
| # (white) | No or negligible ESP |
| (empty) | Not assessed |

**Confidence in evidence**

| | |
|---|---|
| 3 | UK-related, peer-reviewed iterature |
| 2 | Grey or overseas literature |
| 1 | Expert opinion or Obvious |
| (empty) | Not assessed |

**Feature type†**

| | |
|---|---|
| S | Scottish MPA search feature |
| E | English MCZ feature |
| W | Welsh HP MCZ feature |
| NI | Northern Ireland MCZ feature |
| EU | EU Habitats Directive Annex 1 feature or sub-feature |

captured through scoring goods and benefits and inclusion of further columns would reduce the clarity and manageability of the matrices. The framework illustrates obvious linkages between some intermediate regulatory services and final goods and benefits (e.g. between carbon sequestration and healthy climate) although the flow of benefit may depend on the complementary capital invested (i.e. the provision of a significant regulatory service may not necessarily infer a significant benefit for human wellbeing).

Population of the matrices was initially derived from a literature review conducted for the English MCZ process (Fletcher et al. 2012) and was significantly reviewed and updated as part of the Valuing Nature Network (VNN) project (Potts et al. 2014). In particular, the ecosystem services provided by features protected under the EU Habitats Directive (i.e. through SACs) and the additional features proposed for protection by devolved administrations in England, Wales, Scotland and Northern Ireland were added. Since Potts et al. (2014), the framework of ecosystem services has been reviewed as part of the VNN project, the matrices were reviewed and discussed at a national conference in the UK and, as a result of this, additional effort was given to understanding the human health aspects provided by protected features including expert review and discussion.

The list of features in Table 9.1 is separated into broad-scale habitats (where the aim is to ensure adequate representation throughout the UK MPA network) and specific habitats of conservation importance. Bird species, both those protected through the EU Birds Directive and proposed new MPAs, have been excluded in order to contain the manageability of this study.

The shading of each cell represents the relative importance of each feature in providing the respective ecosystem services (darker being more important, lighter less important). Some features are more important than others in providing a particular ecosystem service and therefore scores should be interpreted within the context of all the various features. For example, whilst a number of marine habitats may contribute a climate regulation service, the most important habitats are ‘coastal saltmarshes and saline reedbeds’ and ‘intertidal sediments dominated by aquatic angiosperms’.

The number within each cell relates to the level of confidence that we have in the evidence. Where there was scientific, UK-relevant, peer-reviewed evidence establishing a link between a feature and an ecosystem service, the level of confidence was rated 3. Such a rating is supported by lists of relevant literature and academic papers in Fletcher et al. (2012). A confidence level of 2 indicated support from non-peer reviewed grey literature or overseas literature that was not specifically relevant to either the UK or the particular species (e.g. a closely related species). Where the evidence was based on expert opinion or that of the wider VNN group, then this was given a confidence rating of 1. This also applies to findings that are so obvious there are unlikely to be published papers to such an effect, for example few species will contribute significantly to water cycling. Cells left unnumbered (and uncoloured) reflect gaps in our current understanding. Examination of specific results is undertaken in the Discussion section at the end of this chapter.

## 9.5 How May Site Management Affect the Output of Ecosystem Services from MPAs?

The design and designation of a network of MPAs is only the first step and a rather abstract one. The real effort in conservation comes from the practical management measures put in place to achieve conservation objectives and the enforcement of these measures. Traditionally, MPA management in the UK comes from European designations or small isolated sites, where management measures are determined by the conservation objectives of the site, the sensitivities of the features, and the risks to those features from anthropogenic activities. Such existing sites provide evidence of the impact of management measures on the provision of ecosystem services from MPAs.

For example, in the Moray Firth SAC, protected sub-tidal sandbanks contribute to the delivery of a range of ecosystem services including supporting a number of species of algae and invertebrates; provide protection against erosion; help in maintaining sediment balance; provide minerals for extraction; and spawning grounds and nursery areas for sandeels and juvenile fish, many of which are commercially fished species. This productivity forms an important food source for marine mammals and sea birds which offer cultural services via tourism, nature watching and education. The monitoring evidence to date suggests that the SAC management plan is meeting its objectives for the sandbanks. Ecosystem services, as a consequence, are also being maintained. The status of bottlenose dolphins, also protected by the SAC, is improving and as a consequence it is likely that the considerable tourism services that they support will also be maintained or improved.

Monitoring of the No-Take-Zone off the Isle of Lundy in the Bristol Channel has shown that, as a result of ceasing fishing activities, there is a potential benefit for the surrounding European lobster population through a spill-over of adults and larvae (Hoskin et al. 2009). This is currently being investigated by the Devon and Severn IFCA through tagging surveys and genetic studies of local lobster populations. Evidence on connectivity will be helpful in illustrating the ecological coherence of the MPA network. Other management measures via local byelaws and a zoning plan may further protect the condition of the Lundy marine area, resulting in improved quality of diving at the site and an increase in wildlife visits, thereby improving tourism services. Lundy is further discussed with respect to ecosystem service indicators in Chap. 5.

In the Skomer Marine Nature Reserve (MNR) off the Pembrokeshire coast in Wales, strong evidence points to an increase in the provision of food, tourism and nature watching (diving, site-seeing, nature watching and angling) as a result of existing management plans and measures which precludes some forms of fishing. For example, the scallop population has increased 'at least four fold and perhaps more than eight fold' over the first 20 years of its designation (CCW Press Release, 20 April 2010). Research activities are significant, largely due to the protection and management of the site, and wider educational interest is evidenced by the site hosting visits of 'popular' television programming. Skomer MNR is further discussed with respect to ecosystem service indicators in Chap. 5.

The Lyme Bays reefs, off the Devon and Dorset counties in southern England, are protected from bottom towed fishing gear through a Statutory Instrument: the Lyme Bay Designated Area (Fishing Restrictions) Order 2008. This was driven in part by consideration of supporting service provision, namely formation of species habitat (for both commercially and culturally important species). However, evidence for this, and other functional roles are scarce. Rees et al. (2012) linked supporting ecosystem services (nutrient cycling, bioremediation of waste and gas and climate regulation) with relevant ecosystem processes (energy fixation and transfer, burial, enhancement of microbial decomposition) and assigned them to benthic organisms mapped within the bay using biological traits analysis. The study showed that whilst MPA planning in the UK focuses on protecting specific marine species and habitats, ecosystem services do not map neatly at the species level. Thus, unless key ecosystem service providers are scarce or threatened they are unlikely to be the focus of MPA designation.

## 9.6 Discussion and Conclusions

### *9.6.1 How Are Ecosystem Services Concepts Built into International and UK Policy Relating to MPAs?*

Ecosystem services concepts are inherent in the design concepts of MPA networks under the OSPAR Guidelines, however, they are not directly acknowledged. While MPAs are predominantly designed for conservation, there is growing discussion and practice that MPAs can (and should) deliver increases in human welfare. For example, while the concept of multiple-use has often been contested, it represents a commitment to explore human welfare questions in the adaptive management of marine systems (Agardy et al. 2003). Halpern et al. (2010) acknowledged the need for more studies that linked ecosystem-based management (of which ecosystem services delivery is a part) and the designation of MPAs.

The main priority for MPA policy in the UK is the conservation and recovery of important habitats and species, usually those that are considered nationally important, endangered, threatened or rare. While ecosystem services concepts are not absent, they could be considered supplementary to the existing process. Reference to ecosystem services is sporadic and inconsistent across the different devolved administrations. In England, there is evidence of habitats having been selected for service provision in terms of fisheries recruitment or biodiversity. In the Scottish context there are guidelines that can allow for the selection of sites based on features that provide ecosystem services that underpin human activities. However to date, there are limited examples of actual sites being progressed purely due to ecosystem services in any jurisdiction. While some of the conceptual thinking is there, the reality is that few MPAs are being designated on the basis of identified ecosystem services. Furthermore, the various processes for identifying MCZs in the UK are disjointed both spatially and temporally across administrations with the ecosystem service benefits of sites assessed on an individual basis, preventing a consideration of synergistic effects.

The availability of evidence on ecosystem services has clearly been an obstacle to the implementation of marine biodiversity policy with several rMCZs in England dropped and the Welsh MCZ programme readdressed due to a lack of the perceived benefits of MPAs compared to their socio-economic cost. As is shown by the two matrices developed for this chapter, there are difficulties in providing comprehensive ecological evidence in the UK (and the same is true internationally) about service flows from specific habitats (Herbert et al. 2012) and, hence, a lack of information to support the designation of sites on this basis. Lack of such evidence undermines the designation of sites in terms of their service flows and formation of guidance for policy implementation. Interestingly, the Scottish strategy acknowledges that public support may hinge on the flow of benefits from MPAs to human systems in the short, medium and long term and policy makers appear to recognise that the ecosystem services concept is important in the MPA management process. Therefore, building the database on ecosystem services flows from MPAs is imperative for informing management practice and community support in the long term.

The decisions on the shape of the UK network are not yet fully resolved, with the process at different stages throughout the UK and final designations likely to be spread over several years. Once the UK network is in place, it will be critical to monitor not only the status of listed features, but also the flow of regulating, provisioning and cultural services and benefits from sites and the influence of management instruments on identified services. The way that the pressures within MPAs are managed will determine the scale and type of flows from them and how they relate to areas outside the network. While management plans for the formative UK network will not be negotiated for some months, understanding the pressures and ecosystem service flows will influence the type of management response in the MPA (illustrated by Herbert et al. 2012). We discuss some of these ramifications from the case study sites below.

As the number of MPA designations grows, system-wide benefits to communities from improvements in delivery of a range of services may be realised, but the rate of this will relate to the extent that different ecosystem services are considered when identifying features for protection.

### 9.6.2 What Ecosystem Services May Protected Features in the UK Provide?

What is immediately apparent from the matrices in Tables 9.1 and 9.2 is the significant contribution that new marine species and habitats in the proposed MPA network could potentially make to the existing European network. The habitats listed in Annex I of the Habitats and Species Directive were those that were in danger of disappearance in their natural range, had a small natural range, or presented outstanding examples of typical characteristics of large biogeographic regions (e.g. the Atlantic). The species in Annex II were selected for being endangered, vulnerable, rare or endemic. The focus was largely terrestrial with a relatively small proportion of marine features (13 habitats and 8 species). However, it is worth noting that many of the Annex I habitats are large geographical features, such as estuaries and large shallow inlets and bays, that may comprise a complex of the 'new' habitats and species in Tables 9.1 and 9.2.

The matrices further illustrate that the greatest ecosystem service benefits may be derived from the protection of broad-scale habitats and widespread habitat FOCI such as (Table 9.1), compared to low or limited mobility species (Table 9.2). Broad-scale habitats do not tend to directly provide provisioning goods and benefits, rather they indirectly maintain them to varying degrees through supporting and regulatory services. However, this provision is often specifically related to a particular component or quality of the habitat. For example, the main evidence for nutrient cycling from intertidal rock was related to microbial films; intertidal sediments may support natural hazard regulation where they form natural barriers such as sand banks. Similarly, the formation of species habitats from intertidal rock will be strongly dependent on the nature (composition and complexity) of the substratum itself. As a consequence, it was easier to identify and score, with a greater level of confidence, the more specific and detailed habitat FOCI than the more generic broad-scale habitats.

All habitats may be considered to contribute to the intermediate ecosystem service of 'species habitat' hence a minimum importance score of 'low contribution' is given, but the score may be higher where the habitat or species supports the diversity of other species, for example, through providing a complex microcosm or facilitating settlement and growth. For example, in Table 9.1, maerl beds, horse mussels and seagrass all contribute substantially to habitat complexity.

Table 9.1 identifies the incidence of multifunctional habitats where broad-scale or specific features provide supporting, regulating, provisioning and cultural services across the intermediate and final ecosystem service categories. These systems are highly productive, visible, and coastal and are usually attributed with the best knowledge base as a result of studies published in the peer-reviewed literature. Eight habitat assemblages are apparent from the data. They include broadly defined intertidal systems; coastal salt marshes; intertidal sediments dominated by aquatic angiosperms; subtidal macrophyte-dominated sediment; low or variable salinity habitats; seagrass beds; sea loch egg wrack beds; kelp and seaweed communities on sublittoral sediment; and tide-swept algal communities. These multifunctional habitats are important for the management of MPAs in that they are productive systems providing a diversity of ecosystem services flows. If habitats are to be afforded a priority for conservation other than scarcity or status, it could conceivably be along the lines of diversity or intensity of ecosystem services provision. This prioritisation would have the potential to influence the range of management measures deployed within MPAs, with stricter measures intended for MPAs that produce a wide range of benefits for society.

Some goods and benefits are only provided by particular species, rather than habitats as a whole, such as ornaments, aquaria, medicines and blue biotechnology. Our knowledge of the contribution of especially rare species to ecosystem services or benefits is limited and confined to expert opinion. What is apparent from the data is that certain species play key roles in supporting, provisioning and cultural services but rarely does a species play a dominant role across all types of ecosystem services and benefits.

What is particularly clear is that many species that are considered charismatic play an important role in providing cultural services including spiritual and cultural wellbeing, and tourism and nature watching. Iconic species for example, were particularly relevant for the delivery of psychological benefits (e.g. Atlantic salmon, bottle nose dolphin, basking shark, short and long nosed seahorse and the common skate amongst others). Through this methodology, further research can be explored on the spatial extension and intensity of cultural ecosystem services (CES); a highly relevant outcome for informing policy and engaging communities. It is likely that the social importance of these animals is a consideration for why they are present in MPA designation processes. However, all habitats and species, in the context of seascape, are considered to provide a sense of spiritual and cultural wellbeing, aesthetic benefits, physical and psychological and contribute to education and research to varying degrees.

In terms of the confidence in the habitat assessments, while our understanding derives predominantly from expert opinion (rated 1 in Tables 9.1 and 9.2) and grey/international literature (rated 2), there is reasonable understanding of contributions to ecosystem services, particularly at the scale of intermediate supporting and regulatory services. At a species level, evidence from UK-relevant, peer-reviewed research is considerably lower and there are significant gaps in our knowledge. CES knowledge is patchy and reflects an emerging field of enquiry concerning the identification, spatial extent and valuation of these services. Another dimension in CES that warrants attention is the provision of services from specific habitats and species and the interpretation of CES by individuals and communities. Distinguishing the delivery of CES between different habitat types and the role of broader landscape and seascape in CES delivery is a promising area of inquiry. Furthermore, differentiating between individual values and collective values requires deliberation techniques with the public, and the matrices provide an important starting point for discussion.

How well ecosystem service provision is protected will depend on the designated features selected for each site, and the conservation objectives for those features (to either maintain or recover favourable condition). These objectives focus on the area/population size and quality of the habitat, and not directly on ecosystem services provision.

### 9.6.3 *How May Site Management Affect the Output of Ecosystem Services from MPAs?*

Evidence from a number of existing MPAs in the UK illustrates the potential ecosystem service benefits that may arise from appropriate management measures and effective management plans. Despite this evidence, it remains difficult to confidently predict the outcomes of specific management measures on specific features in a specific location. The ability to then predict improvements in the provision of ecosystem services by MPA management is limited by our understanding of complex ecological interactions such as trophic cascades. Furthermore, factors outside of an MPA boundary may influence the delivery of conservation objectives, such as land management issues and climate change. Such a range of factors from complex internal site interactions to complex external site processes mean that we will rarely be able to confidently predict the patterns of ecological change brought about by the implementation of protection measures in an MPA.

Herbert et al. (2012) compiled evidence on the success rates of more than 40 management measures in relation to the conservation of specific features and attempted to identify the factors that influenced their success. Whilst there was good evidence for impacts on features, there was less evidence for impacts on ecosystem services. The evidence base would be improved if monitoring programmes aimed to quantify the impacts of measures on the ecology, ecosystem processes and ecosystem services of habitats and features.

However, given inherent unpredictability in ecological systems, the large cost involved in gathering relevant scientific data to support an evidence-based policy approach, and considering the potentially greater cost of not protecting our seas, decisions on the designation of MPAs may need to be pragmatic in the assessment of 'sufficient evidence'.

### 9.6.4 *International Context*

One definition of an MPA by the International Union for Conservation of Nature is a geographical area of land and water that is recognised, dedicated and managed through legal or other effective means to achieve the long-term conservation of nature with associated ecosystem services and cultural values (after Dudley 2008). Design guidelines in other countries refer explicitly to the role that MPAs can play in delivering ecosystem services, in particular, ecosystem resilience. Guidelines prepared for the North American MPA Network identify that, if MPA networks are focussed on sustaining key ecological functions, services and resources, they can help to mitigate climate change effects on ecosystems (Brock et al. 2012). Integrating an ecosystem approach into conservation planning recognises that biodiversity is not static in time or space but is generated and maintained by dynamic natural processes (Pressey et al. 2007).

## 9.7 Conclusions

Ecosystem services provide a useful bridge between natural and social sciences for recognising and valuing the benefits that humans obtain from healthy functioning coastal and marine systems. This chapter highlights that while the data on identifying and valuing ecosystem services flows is incomplete and at an early stage of collection, the concept is important in understanding our relationship to coastal systems and the benefits of conservation and protection.

An investigation of ecosystem service flows often elucidates wider ecosystem processes and linkages that operate outside of discrete site boundaries and enables a comprehensive assessment of these. In particular, this may highlight the influence of external pressures such as land management practices and climate change on protected features across the network of sites.

The matrices of marine features identified for protection in UK waters illustrate the greater diversity and flow of ecosystem services provided by broad-scale features as compared to the rare and threatened habitats and species that are so often the focus on conservation efforts. It is important that broad-scale features are also protected by management measures within the new MPA process in order to maximise the potential for benefits from ecosystem services.

In terms of MPAs, few designation processes have explicitly taken the ecosystem services concept into account despite the recognition in policy of its importance. We argue that this is primarily due to a lack of information and guidance rather than explicit omission, and that future designation and management of MPAs should take into account the flow of ecosystem services from protected marine habitats to human systems across the portfolio of supporting, regulatory, provisioning and cultural services and their associated goods and benefits. Integrating such design concepts into marine conservation planning will better protect marine features that are themselves dynamic and also needing to survive and adapt in a highly dynamic environment.

This approach would complement ongoing developments under the MSFD where ecosystem service assessments have been carried out to assess the impacts of potential management measures needed to deliver Good Environmental Status. The draft programme of measures may include MPAs (both existing and proposed) and will be consulted upon in autumn 2014.

## References

Agardy, T., Bridgewater, P., Crosby, M. P., Day, J., Dayton, P. K., Kenchington, R., Laffoley, D., McConney, P., Murray, P. A., Parks, J. A., & Peau, L. (2003). Dangerous targets? Unresolved issues and ideological clashes around marine protected areas. *Aquatic Conservation: Marine and Freshwater Ecosystems, 13*, 353–367.

Brock, R. J., Kenchington, E., & Martínez-Arroyo, A. (2012). *Scientific guidelines for designing Resilient marine protected area networks in a changing climate*. Montreal: Commission for Environmental Cooperation. 95pp.

Defra. (2012). *Marine conservation zones: Consultation on proposals for designation in 2013 December 2012*. Department of Environment, Food and Rural Affairs. Available from: http://www.defra.gov.uk/consult/2012/12/13/marine-conservation-zones-1212/

DoE. (2013a). *Draft guidance on selection and designation of Marine Conservation Zones (MCZs) in Northern Ireland Inshore Region*. Belfast: Department of the Environment.

DoE. (2013b). *A draft strategy for marine protected areas in the Northern Ireland Inshore Region*. Belfast: Department of the Environment.

Dudley, N. (2008). *Guidelines for applying protected area management categories*. Gland: IUCN. 86pp.

Earnshaw, S., Perez-Dominguez, R., Worsfold, T., Pears, S., Hubble, M., Blyth-Skyrme, R., Wells, E., Blyth-Skyrme, V., Hall-Spencer, J., & Rehfisch, M. (2014). Independent science review of the Scottish marine protected areas network. APEM Scientific Report for The Scottish Government, Report Ref: APM/002/14. APEM Ref 413361.

Fletcher, S., Saunders, J., Herbert, R., Roberts, C., & Dawson, K. (2012). *Description of the ecosystem services provided by broad-scale habitats and features of conservation importance that are likely to be protected by Marine Protected Areas in the Marine Conservation Zone Project area*. Natural England Commissioned Reports, Number 088.

Folke, C., Carpenter, S., Walker, B., Scheffer, M., Elmqvist, T., Gunderson, L., & Holling, C. (2004). Regime shifts, resilience, and biodiversity in ecosystem management. *Annual Review of Ecology, Evolution, and Systematics, 35*, 557–581.

Halpern, B. S., Lester, S. E., & McLeod, K. L. (2010). Placing marine protected areas onto the ecosystem-based management seascape. *Proceedings of the National Academy of Sciences, 107*(43), 18312–18317.

Herbert, R., Saunders, J., Bowles, A., McKinley, E., Pegg, J., Pearson, A., Villasmil, I., & Fletcher, S. (2012). *Potential impacts of management measures on the ecology and ecosystem services of Marine Protected Areas in UK waters*. Report to the Joint Nature Conservation Committee, Peterborough. ISSN 0963 8901.

Hoskin, M. G., Coleman, R. A., & von Carlshausen, L. (2009). *Ecological effects of the Lundy No-Take Zone: The first five years (2003–2007)*. Report to Natural England, DEFRA and WWF-UK.

Jackson, E.L., Hiscock, K., Evans, J., Seeley, B., & Lear, D. (2008). *Investigating the existing coverage and subsequent gaps in protection and providing guidance on representativity and replication for a coherent network of Marine Protected Areas in England's territorial waters*. Plymouth: Marine Life Information Network (MarLIN), Marine Biological Association of the UK. Natural England Commissioned Reports, Number 018.

JNCC and Natural England. (2010). *Ecological network guidance*, Peterborough. 124pp.

Loreau, M., Naeem, S., Inchausti, P., Bengtsson, J., Grime, J. P., Hector, A., Hooper, D. U., Huston, M. A., Raffaelli, D., Schmid, B., Tilman, D., & Wardle, D. A. (2001). Biodiversity and ecosystem functioning: current knowledge and future challenges. *Science, 294*, 804–808.

Marine Scotland. (2011). *A strategy for marine nature conservation in Scotland's seas*. Scottish Government. Available from: http://www.scotland.gov.uk/Topics/marine/marine-environment/Conservationstrategy/marineconstrategy/

MEA. (2005). *Millennium ecosystem assessment – Ecosystems and human wellbeing biodiversity synthesis*. Washington, DC: Island Press.

Natural England and the Joint Nature Conservation Committee. (2010). *Marine conservation zone project: Ecological network guidance*. Sheffield/Peterborough. Available from: http://jncc.defra.gov.uk/pdf/100705_ENG_v10.pdf

Olds, A. D., Pitt, K. A., Maxwell, P. S., & Connolly, R. M. (2012). Synergistic effects of reserves and connectivity on ecological resilience. *Journal of Applied Ecology, 49*, 1195–1203.

OSPAR. (2006). *Guidance on developing an ecological coherent network of OSPAR marine protected areas*. Reference number 2006–3. Available from: http://www.ospar.org/content/content.asp?menu=00180302000011_000000_000000

Palumbi, S. R. (2003). Population genetics, demographic connectivity, and the design of marine reserves. *Ecological Applications, 13*, 146–158.

Potts, T., Burdon, D., Jackson, E., Atkins, J., Saunders, J., Hastings, E., & Langmead, O. (2014). Do marine protected areas deliver flows of ecosystem services to support human welfare? *Marine Policy, 44*, 139–148.

Pressey, R. L., Cabeza, M., Watts, M. E., Cowling, R. M., & Wilson, K. A. (2007). Conservation planning in a changing world. *Trends in Ecology & Evolution, 22*(11), 583–592.

Rees, S. E., Austen, M. C., Attrill, M. J., & Rodwell, L. D. (2012). Incorporating indirect ecosystem services into marine protected area planning and management. *International Journal of Biodiversity Science, Ecosystem Services & Management, 8*, 272–285.

Roberts, C., Hawkins, J., Fletcher, J., Hands, S., Raab, K., & Ward, S. (2010). *Guidance on the size and spacing of marine protected areas in England.* York: Environment Department University of York.

Roncin, N., Alban, F., Charbonnel, E., Crec'hriou, R., De la Cruz Modino, R., Culioli, J.-M., Dimech, M., Goñi, R., Guala, J., Higgins, R., Lavisse, E., Direach, l. L., Luna, B., Marcos, C., Maynou, F., Pascual, J., Person, J., Smith, P., Stobart, B., Szelianszky, E., Valle, C., Vaselli, S., & Boncoeur, J. (2008). Uses of ecosystem services provided by MPAs: How much do they impact the local economy? A southern Europe perspective. *Journal for Nature Conservation, 16*, 256–270.

Rondinini, C. (2011). *Meeting the MPA network design principles of representation and adequacy – developing species-area curves for habitats.* JNCC Report 439, ISSN 0963–8091.

Scottish Government. (2012). *Report to the Scottish parliament on progress identify a Scottish network of marine protected areas.* Available from: http://www.scotland.gov.uk/Topics/marine/marine-environment/mpanetwork/MPAParliamentReport

# Chapter 10
# Why Value 'Blue Carbon'?

**Tiziana Luisetti, R.K. Turner, Julian Andrews, Emma Jackson, Maria Giovanna Palmieri, Antara Sen, and Lucille Paltriguera**

## 10.1 Introduction

Coastal 'blue' carbon ecosystems (mangroves, saltmarshes, and seagrasses) are important carbon storage providers. Of all the "green" carbon captured by photosynthetic activity in the world over half (55 %) is captured by marine living organisms (Nellemann et al. 2009) and for this reason is called "blue" carbon. According to Mcleod et al. (2011) and Sifleet et al. (2011), saltmarshes and mangroves (both oceanic and estuarine) store more carbon per hectare than tropical forests. Terrestrial carbon storage providers (e.g. tropical forests) are protected by international

T. Luisetti (✉) • L. Paltriguera
Centre for Environment, Fisheries & Aquaculture Science (CEFAS),
Pakefield Road, Lowestoft, Suffolk NR33 0HT, UK
e-mail: tiziana.luisetti@cefas.co.uk

R.K. Turner
School of Environmental Sciences, University of East Anglia, Norwich, UK

J. Andrews
School of Environmental Sciences, University of East Anglia, Norwich,
Norfolk NR4 7TJ, UK

E. Jackson
School of Medical and Applied Sciences, Central Queensland University,
Bryan Jordan Drive, PO Box 1319, Gladstone, QLD 4680, Australia

M.G. Palmieri
Centre for Social and Economic Research on the Global Environment (CSERGE),
School of Environmental Sciences, University of East Anglia,
Norwich Research Park, Norwich NR4 7TJ, UK

A. Sen
Affiliate of the Centre for Social and Economic Research on the Global Environment
(CSERGE), School of Environmental Sciences, University of East Anglia,
Norwich Research Park, Norwich NR4 7TJ, UK



R.K. Turner, M. Schaafsma (eds.), *Coastal Zones Ecosystem Services*,
Studies in Ecological Economics 9, DOI 10.1007/978-3-319-17214-9_10

mechanisms such as REDD+ (an incentive mechanism proposed by the United Nations Framework Convention on Climate Change (UNFCCC) at the Conference of Parties held in Montreal in 2005. In this scheme, developed countries financially compensate developing countries to reduce their emissions from forests through reduced deforestation and degradation, and sustainable forest management). But similar international mechanisms are currently not available for blue carbon ecosystems (da Silva Copertino 2011; Fourqurean et al. 2012). Nevertheless, about half of the blue carbon ecosystems have been lost in the last half century (Thomas 2014). Therefore, the potential for including blue carbon into existing global climate policy mechanisms has been currently explored and diverse economic incentives investigated for their suitability in the 'blue' ecosystem context (Murray et al. 2011).

Thomas (2014) presents probably the first comprehensive literature review of blue carbon which has emerged largely since 2011. Thomas analysed 46 documents, including peer-reviewed papers and 'grey literature', published between 2009 and 2013. The results of the literature review show that most of these documents are either about the science or the economics of blue carbon. The science looks at the regulating service of blue carbon and the ecosystems providing it. Biophysical data about the blue carbon budget in mangroves, saltmarshes and seagrasses are still limited and uncertainties remain high. The economics of blue carbon literature explores the value of this service with the aim of supporting the preservation of the marine and coastal ecosystems that provide blue carbon and other co-benefits (e.g. flood protection, biodiversity, nursery habitats, and improved water quality). Thomas' review highlights that for the management of coastal blue carbon to be successful, appropriate stakeholder involvement through local communities is crucial. However, the current blue carbon literature does not contain any detailed social study of such deliberation processes.

According to Thomas (2014), current blue carbon literature identifies two main mechanisms that could be potentially used to successfully enhance the protection of blue carbon and related ecosystems: an extension of REDD+ to coastal areas, and the development of payment for ecosystem services (PES) in coastal areas. However, it has been argued (Gómez-Baggethun et al. 2010) that there is the risk that PES mechanisms might encourage resource exploitation when financial returns are insufficient. This is an example of the possible negative effects of the so called 'commodification of nature' phenomenon. Furthermore, since the price of trading blue carbon can be a lot lower than the returns from shrimp farming, Vandergeest et al. (2009) suggest that blue carbon payments are a more feasible option for those operating at the marginal levels of financial sustainability in developing countries as they are less likely to have the resources to invest in commercial shrimp farming. Other actions taken by nation-states in developing countries in line with their commitments under the UNFCCC are the Nationally Appropriate Mitigation Actions (NAMAs) – with these actions the benefits of carbon accrue to the implementing nation. In contrast, REDD+ operates between nation-states, with financial flows from a developed country to a developing country; the first will gain carbon offset benefits to meet its own commitments, the latter sustainable development benefits, and globally all parties may benefit from a healthier climate and conserved biodiversity.

Some recent studies have investigated national and global estimates of emissions with a focus on the value of the impact caused by the destruction or degradation of blue carbon ecosystems. An added bonus has been a rising of the blue carbon profile as suggested by Ullman et al. (2013). These studies attempted to value blue carbon ecosystems by estimating: the total damage cost of the release of carbon from global blue carbon habitats (Pendleton et al. 2012); the value of UK blue carbon habitats to sequester and store carbon (Beaumont et al. 2014), and also tried to clarify the valuation issues around the 'stock' and 'flows' of carbon in European blue carbon habitats (Luisetti et al. 2013).

Many ecosystem services such as carbon sequestration and storage as well as the other co-benefits provided by blue carbon ecosystems can be meaningfully expressed in monetary terms. This chapter aims to raise awareness in the political domain about the 'blue' carbon issue through this type of economic calculus which can be used to further conservation efforts. We review two coastal blue carbon valuation case studies in the UK (Luisetti et al. 2014) and at the European level (Luisetti et al. 2013). With the aid of scenario analysis, these studies assess how human welfare benefits may be affected by changes in coastal blue carbon provision. Although there is evidence that carbon sequestration capacities and the full development of other ecosystem services can be lower in restored coastal habitats (Matsui et al. 2012; Mossman et al. 2012), these case studies show that in appropriate contexts restoration/re-creation schemes such as managed realignment practice can provide net economic benefits.

## 10.2 Carbon Stock and Flows: An Analysis of the Organic Carbon Cycling, Burial, and Storage in the Temperate Estuarine Mudflat-Saltmarsh System

Blue carbon ecosystem stocks (the coastal and marine ecosystem structure, processes and functions) possess high biological productivity (McLusky and Elliott 2004) and provide a diverse set of habitats and species, with a consequent flow of ecosystem services of significant benefit (value) to society (Fisher et al. 2009). As reported by Luisetti et al. (2014) organic carbon input and cycling within the estuarine water column is itself complex (Boyes and Elliott 2006) and this complexity is maintained as organic carbon is deposited in estuarine sediments. In this section a summary of the complex carbon cycling in the temperate estuarine mudflat-saltmarsh system is presented (see Luisetti et al. 2014 for a more detailed explanation). Organic matter typically arrives in sediments either as coatings on minerogenic particles, or as discrete particles (Andrews et al. 2011). Once deposited in sediment, organic matter is typically subject to a series of aerobic and anaerobic microbial decay mechanisms (e.g. Bianchi 2007). Overall these decay mechanisms can serve as an intermediate ecosystem service; for example the microbially-driven denitrification reaction uses the dissolved nutrient compound nitrate in seawater to help oxidise organic matter (e.g. see Bianchi 2007), a reaction that consumes the nitrate, acting as a natural 'waste treatment' process, that helps improve water quality (Seitzinger et al. 2006; Andrews et al. 2006; Turner et al. 2008). However, this reaction also produces a small but significant quantity of the potent greenhouse gas nitrous oxide (Andrews et al. 2006; Middelburg and Levin 2009; Jickells and Weston 2011; Adams et al. 2012), which through its role in global warming is a dis-benefit rather than 'a service'.

The residual organic carbon burial becomes semi-permanent and classed as storage (Andrews et al. 2011). This net organic carbon storage (final ecosystem service) can be directly measured (e.g. Andrews et al. 2008; Sousa et al. 2010) and increases as long as there is sediment available for deposition and the carbon supply to that sediment is maintained. Storage (a service flow over time) in estuarine mudflat-saltmarsh systems (the stock) operates on timescales of decades (Andrews et al. 2008) and centuries (Allen and Rae 1987) to millennia (Rees et al. 2000; Andrews et al. 2000); it can be viewed as a societal benefit in terms of an atmospheric $CO_2$ equivalent that has been removed from the atmosphere, thus reducing greenhouse gas impacts. This final ecosystem service becomes more valuable as time passes, which implies that we should protect, and even enable deposition of muddy estuarine accumulation (Turner et al. 2007; Luisetti et al. 2011).

## 10.3 Carbon Stock and Flows: The Debate Around Their Economic Valuation

The debate around the economic valuation of stock and flows has been recently analysed in the literature looking at the carbon cycle in coastal blue carbon ecosystems (Luisetti et al. 2013, 2014). From an evaluation point of view, it is in fact possible to consider the economic value of the flow of benefits provided by coastal and marine ecosystems and the accounting value of the current ecosystem stock.

The ecosystem stock price value is estimated, in accounting terms, at a precise point in time (Banzhaf and Boyd 2012; Costanza et al. 2014) and the economic valuation of flows is related to marginal changes in the stock over time. In the 1990s, Costanza et al. (1997) produced a global ecosystem service value estimation. This work was subsequently critiqued on a number of grounds including that the aggregate value was not necessarily the sum of the parts, and that US$33 trillion (the Costanza et al. calculation) was more than global income and therefore peoples' ability to pay (Heal 2000). Further work (Howarth and Farber 2002) sought to defend the Costanza et al. approach by arguing that the estimates of ecosystem services value were analogous to National Income Accounting entities such as GDP with a constant set of value weights. The underlying rationale is that the aggregate measure is a quantity parameter (the stock concept), and, while it is related to value, it does not directly value the planet's ecosystem services in total. In this sense it is an accounting price measure of the quantity of ecosystem services holding prices constant, where the measures are not based on economic theory but on accounting rules. The current extent of European coastal blue carbon (the carbon storage service provided by saltmarshes and sea grasses) has, for example, an accounting stock price (value) of about US$180 million, valued at EUA price[1] (Luisetti et al. 2013). Such total (stock) values can be estimated and compared for two different points in time as a heuristic to help to appreciate the change in natural capital. This viewpoint is, however, controversial and is not supported by many mainstream economists. However, Luisetti et al. (2014) argue that estimates of aggregate stock accounting price value can play a valuable role in the 'politics' of the environment and can heighten awareness of the overall importance of ecosystem services relative to and in combination with other contributors to human well-being.

For economic valuation, however, it is important to be able to quantify and evaluate gains or losses in stock assets and consequent service flows. In this case, the focus is on determining the 'marginal economic value' as it relates to an incremental increase/decrease in a set of ecosystem services over time and space. Neither the stock value nor the marginal flow value for ecosystem services in themselves provide warning signals about possible threshold effects leading to ecosystem collapse. The so called total economic value supplied by an ecosystem's set of final services is always less than the total system value (Turner et al. 2003) and a precautionary attitude in conservation policy and adaptive management seems prudent. But economists would also argue that any precautionary measure such as safe minimum standards is not a silver bullet. Society will still have to face up to trade off decisions and judgements about the cost acceptability (opportunity cost) of precautionary measures. Another important consideration is that the flow economic values and the stock accounting price values cannot be aggregated. They are, heuristically speaking, complements in the sense that they provide different dimensions of, or perspectives on, the magnitude and significance of ecosystem services.

## 10.4 Carbon Stock and Flows: A Valuation Framework and Two Case Studies

Storage of carbon, and thus the reduction of carbon dioxide in the atmosphere, is regarded as a valuable ecosystem service. For terrestrial systems research has been conducted on valuing this service. Here we review an equivalent valuation framework adopted for coastal blue carbon systems.

To value the changes (gain/losses) in carbon storage capacity associated with increased or decreased coastal ecosystem areas, the case studies extrapolated from Luisetti et al. (2013, 2014) that we review in this chapter make use of scenario analysis to analyse any changes that may occur in plausible futures (usually based on available climate change model projections); more 'standard' as well as more

[1] EU Allowance (EUA) is a mean price of traded carbon in the European Union emissions trading scheme (EU ETS), which is the first and the largest international system for trading greenhouse gas emission allowances in operation.

extreme scenarios are considered. The main data required within a scenario analysis for ecosystem services valuation are the following:

1. the (final) ecosystem service (UK National Ecosystem Assessment Follow-on, 2014) under investigation (e.g. carbon sequestration and storage);
2. the ecosystem providing the service of interest (e.g. coastal and marine ecosystems);
3. detailed maps of the extent of those ecosystems;
4. the 'quantity' of the service provided (e.g. carbon burial rates);
5. the accounting and/or the economic value of the service, depending on the purpose of the economic analysis.

In this chapter we are interested in the coastal blue carbon sequestration and storage service. European coastal blue carbon sinks include saltmarsh and seagrass beds only, which represents the globally most important coastal blue carbon storage sink and service benefit in both living and organic-reach soils (Fourqurean et al. 2012).

Luisetti et al. (2013) report that the area extent of European saltmarshes can be estimated from the maps of the European Environment Agency – CORINE land cover maps 2000 and 2006. Although overlaying the CORINE maps to analyse saltmarsh rate of change produces contradictory results, the authors explain that this may simply be an artefact of better mapping techniques in use by 2006 and estimate the current extent of European saltmarshes at 330,653 ha.

Both case studies reviewed in this chapter make use of the Adams et al. (2012) estimates of net carbon sequestration in natural and mature managed realignment saltmarsh located in the Blackwater estuary on the East of England given a 5.4 mm assumed sedimentation rate. To our knowledge, these estimates are currently the only carbon burial net (of the greenhouse gas emissions released back in the atmosphere in the process of carbon burial) estimates available in Europe.

In terms of the blue carbon habitats themselves, *Posidonia oceanica* is an endemic species in the Mediterranean. It is not only the most abundant and widespread seagrass species but also the one that is able to best capture $CO_2$ from the atmosphere (Kennedy and Bjork 2009). In other European seas the most dominant seagrass is the eelgrass, *Z. marina*. Official European mapping estimates of seagrass cover similar to those of saltmarshes are not currently available. Therefore, Luisetti et al. (2013) brought together estimates from various sources (see Luisetti et al. 2013 for details). In total they have estimated the current extent of European seagrasses to be 2.5 million ha of *P. oceanica* and 239,242 ha of *Z. Marina*.

Despite the recognised importance of seagrasses as carbon storage sinks, there have been few studies assessing the carbon burial rates for these ecosystems; for example, knowledge of the sequestration capacity of *Z. marina* beds is rudimentary. As reported by Luisetti et al. (2013), seagrasses have a number of characteristics which result in high carbon burial rates (Cebrian et al. 1997), particularly in terms of slow growing long lived species such as *P. Oceanica*. Accretion rates of carbon also vary from site to site due to currents, growth rates and wave exposure. Furthermore carbon burial rates vary between species of seagrass and the dominant species of seagrass is different in each European sea region.

For the purpose of sensitivity analysis, in the case studies reported in this chapter, the changes of carbon storage service flows in the different scenarios were valued by two different methods: the social cost of carbon (SCC), which measures the monetary value of the avoided carbon releases to the atmosphere because of storage (damage cost avoided method), and the Department of Energy and Climate Change (DECC) prices for non-traded carbon (DECC 2011), based on the marginal abatement cost (clean-up cost) method. In addition the 'market' price (the average EU Allowance spot price) is used to calculate the value of the stock of saltmarshes in terms of carbon sequestration and storage at a specific point in time in both case studies.

### *10.4.1 Case Study: Blue Carbon Valuation in the UK*

This case study is about valuing the carbon sequestration and storage service in the saltmarsh ecosystems of two main estuaries located on the east coast of England: the Humber and the Blackwater estuaries. In this case study, Luisetti et al. (2014) also show the potential for transferability of biophysical and value estimates related to the valuation of the carbon storage service.

The Humber estuary is a relatively large and complex estuary with an industrialised catchment. In the past, the estuary has been subject to major land claim which has denuded the intertidal areas. Most of its intertidal area has been lost (about 90 %), which has greatly reduced the capacity for the storage and processing of carbon, nutrients and of the abundant contaminants in the estuary (Jickells et al. 2000; Andrews et al. 2006; McLusky and Elliott 2004).

The Blackwater has also been subject to land claim and loss of intertidal areas. In contrast to the Humber, the Blackwater is located in a region that is largely semi-rural with a catchment dominated by intensive agriculture with small amounts of industrial legacy contaminants (Shepherd et al. 2007).

These estuaries are close geographically and geomorphically, and agricultural impacts from the catchments are probably broadly comparable (Shepherd et al. 2007). Furthermore, there is empirical evidence that the carbon storage terms for the Southern North Sea tidal fringe are remarkably similar (see Andrews et al. 2011). Intertidal area re-creation, through managed realignment (MR) for example, is considered a priority because of the risks related to sea level rise in both the Humber and the Blackwater (Elliott et al. 2014; Mazik et al. 2010; Edwards and Winn 2006).

These estuaries have been chosen as a case study because, as argued by Luisetti et al. (2014), it is possible that in the context of the next 50–100 years adverse system change is likely to result in both the Humber and the Blackwater estuary. Under a combination of rapid sea-level rise (exacerbated in the English east coast by isostatic adjustments), coupled with climatic change resulting in increased winter rainfall, increased incidence of extreme events and increased storminess of the North Sea the present intertidal area will be drowned and its ecosystem service provision potentially reduced.

In this case study, two scenarios – adapted from previous scenario studies (Langmead et al. 2007; Turner et al. 2007; Luisetti et al. 2011) – and a baseline scenario have been considered over the period 2010–2110. Two managed realignment scenarios: *baseline* or minimum ecosystem services (MinES) scenario; a maximum ecosystem services provision (MaxES) scenario. A net loss of ecosystem services (LESS) scenario is also considered; in this scenario the implementation of the EU Habitats (92/43/EEC) and Birds (2009/147/EC) Directives allowing for saltmarshes re-creation is imagined to become limited, and intertidal habitats continue to be squeezed between the defences and the rising sea level.

Previous studies (Turner et al. 2007; Luisetti et al. 2011) already evaluated managed realignment schemes as a policy response to the loss of intertidal ecosystem services for the Humber and the Blackwater estuary providing some cost-benefit analysis (CBA) results. In this chapter the focus is on the comparison of stock accounting and flow economic values for ecosystem services, and more specifically the regulating service of carbon storage, which provides a healthy climate.

In Luisetti et al. (2014) calculations for the stock and the flow estimates of the ecosystem services provided by the estuaries are based on the extent of saltmarshes estimated for each scenario in previous studies (Luisetti et al. 2011; Turner et al. 2007; Andrews et al. 2000). In the maximum provision scenario, based on the evidence on colonization of managed realignment sites provided by the literature (Dagley 1995; Mossman et al. 2012; Wolters et al. 2008; Morgan and Short 2002), the authors assumed that full provision of ecosystem services will be effective after 5 years from realignment in the Blackwater and after 15 years in the Humber estuary, and that this will be maintained until 2110. The net loss scenario follows the predictions for saltmarsh loss because of sea level rise reported in Jones et al. (2011). Although the authors acknowledge that other evidence may suggest a different percentage loss of saltmarshes for the UK coastline, for illustrative purposes in the case study it was assumed that the loss of saltmarshes will continue at the fixed rate of 0.225 % per annum calculated on the baseline in 2010 till 2110.

The estimates in Adams et al. (2012) for the Blackwater estuary are used as follows: the estimate in managed realignment saltmarshes (1.15 $tC\ ha^{-1}\ year^{-1}$) for the maximum provision (MR) scenario analysis (MaxES), and the estimate in natural saltmarshes (0.94 $tC\ ha^{-1}\ year^{-1}$) for the net loss scenario (LESS), and for the calculation of the current stock of the carbon sequestration service (in this case tC were converted in $tCO_2$).[2]

For the economic valuation, the future flows of services value were discounted to obtain their present value. The present value of the services flow for a time horizon

[2] When attributing a monetary value to an amount of carbon (\$tC) or carbon dioxide (\$$tCO_2$) respectively, the actual carbon content of carbon dioxide has to be taken into account to ensure the "damage cost" is normalised between the two units of measure. $CO_2$ weights 44 g/mol, of which 12 g/mol is the mass of carbon and 32 g/mol the mass of the two oxygen atoms. Therefore the carbon content (and associated value/damage cost) of carbon dioxide is 12/44 (just over 25 %) of the value of pure carbon, or in reverse the value of 1 t of carbon is 44/12 (approximately 4 times) that of 1 t of carbon dioxide. This implies that the monetary value of the damage cost presented in \$tC is equivalent to the damage cost presented in \$$tCO_2$.

**Table 10.1** Present value of carbon sequestration flows for the Blackwater and Humber estuaries over 100 years (2010–2110) discounted following the declining discount rate scheme of the UK Green Book (HMT 2011) under two different scenarios: MaxES; LESS (£*1,000)

| | SCC/tC | | | DECC[a] price/$tCO_2$ |
|---|---|---|---|---|
| | £ 7 | £ 30 | £ 230 | All relevant year values (£) |
| *Blackwater* | | | | |
| (MaxES) scenario | 527 | 2,257 | 36,670 | 45,500 |
| (LESS) scenario | 0.5 | 2 | 15 | 36 |
| *Humber* | | | | |
| (MaxES) scenario | 28 | 123 | 941 | 3,000 |
| (LESS) scenario | 3 | 15 | 117 | 325 |

[a]DECC (2013) provides non-traded C prices till 2100. Here, we assume that the trend showed in the relevant year values between 2096 and 2100 will continue till 2110

of 100 years in each scenario is calculated applying the declining discount rate scheme as recommended by current British policy appraisal (Treasury Green Book) (HMT 2011). Results are reported in Table 10.1.

Finally, the 'market' price (the average EU Allowance spot price in 2010 (€15/$tCO_2$) (Rickels et al. 2010; Chevallier 2010)) was used to calculate the accounting price value of the stock of saltmarshes in terms of C sequestration and storage in the Blackwater (€56,000) and Humber (€575,000) estuaries at a specific point in time.

Luisetti et al. (2014) concluded that the carbon sequestration and storage service is particularly suited to cross-disciplinary analysis. In the case study reviewed here biophysical estimates have been transferred without any adjustment within the two estuaries and the accounting and economic value for carbon storage estimated using carbon prices and global social cost of carbon estimates respectively. This makes its value transfer highly possible within the same country boundaries, and possibly at the European level as well.

## *10.4.2 Case Study: Blue Carbon Valuation in Europe*

This case study is about valuing the carbon sequestration and storage service in European blue carbon ecosystems at the European level. Economic valuation is undertaken for the current stock of blue carbon ecosystems in Europe and for the changes (decrease) of the carbon storage service following loss of saltmarsh and seagrass areas. Luisetti et al. (2013) examined the changes in the coastal blue carbon storage service provision looking at three possible future scenarios: the first (S1) based on the continued application of current coastal ecosystem conservation policies; the second (S2) considered the risk of future benefit losses due to the potential lack of protection for coastal 'blue' carbon ecosystems because of financial budget restrictions, and sea level rise effects; the third (S3) scenario is like the S2 scenario for the saltmarshes, but more *extreme* for seagrasses as here the possible functional (and related services) extinction of *P. oceanica* by 2060 (Forequen et al. 2012) is taken into account.

For saltmarshes, in the first scenario the authors imagined that the rate of loss will still be relatively small for saltmarshes (0.225 % per annum, which is equal to a loss of 4.5 % over 20 years (Jones et al. 2011)); in the second scenario a loss of 0.225 % per annum for the first 20 years and a further loss of 0.3 % per annum for the following 30 years (equal to a loss of 6 % over 20 years – an average between 4.5 % and the projected habitat loss of 8 % by 2060 (Jones et al. 2011)) was assumed. Therefore, in the first scenario 744 ha of European saltmarshes are lost each year, which over 50 years is equal to a loss of 37,200 ha. In the second scenario over 50 years, the loss is equal to 43,290 ha.

In both scenarios, the authors imagine a continuous loss of seagrass beds of 1 ha per day (Langmead et al. 2007). Assuming a future 1 ha per day loss of *P. oceanica* over 50 years (years of 365 days) and that the extent of *P. oceanica* in the Mediterranean is currently about 2,500,000 ha, the amount of *P. oceanica* lost by 2060 will be equal to 18,250 ha. In the system collapse scenario (S3) by 2060 the loss of *P. oceanica* might be 2,250,000 ha (90 % of current extent) (Jorda et al. 2012). That is roughly equal to a loss of 45,000 ha per annum.

As in the previous case study, the Adams et al. (2012) carbon burial estimates for saltmarshes were used in this case study too. The authors considered that managed realignment sites are still experimental and hence still quite limited in Europe. Therefore they used the Adams et al. estimate of natural saltmarshes (0.94 tC $ha^{-1}$ $year^{-1}$) assuming that this estimate is suitable also for the whole north of Europe climatic zone. For the south of Europe, since it is reported in the literature that average annual temperature explains only 5 % of the variability in rates of carbon sequestration, and that there is no significant difference between average rates of carbon sequestration in mangroves and saltmarsh (Chmura et al. 2003), the authors used the same estimate. For the seagrasses the carbon burial rate estimated by Gacia et al. (2002) for *P. oceanica* (1.82 tC $ha^{-1}$ $year^{-1}$) and the rate reported for a Spanish seagrass meadow (Cebrian et al. 1997) for *Z. marina* (0.52 tC $ha^{-1}$ $year^{-1}$) were used.

The accounting price value of the total carbon stock in European saltmarshes and seagrass beds was also estimated. According to Blue next, the average EU Allowance (EUA) price in June 2012 was €8/t$CO_2$. The estimated accounting value in Europe at EUA price was estimated to be US$11,203,843 and US$168,749,727 for current saltmarshes and seagrass beds respectively.

Considering that threshold limits are unknown and given the risk of the extreme scenario in which the potential functional extinction of *P. oceanica* will become a reality in a relatively short period of time, the authors also used marginal values to value increases and/or decreases in coastal blue carbon vegetation. The present value of the flows of carbon storage benefits were discounted at a 3.5 % constant discount rate. Results are reported in Table 10.2.

**Table 10.2** Present value (US$) of the carbon storage service economic value in European blue carbon ecosystems in the optimistic, pessimistic and ultra-pessimistic scenarios over 50 years (2010–2060), discounted at 3.5 % discount rate, at SCC and DECC prices. Conversion rate: £1 = US$1.50 (DECC prices)

| Scenario | Saltmarshes | | Seagrasses | | European blue carbon ecosystems |
|---|---|---|---|---|---|
| | Area lost (ha) | Carbon storage service loss economic value by 2060 (US$) | Area lost (ha) | Carbon storage service loss economic value by 2060 (US$) | Total carbon storage service loss economic value by 2060 (US$) |
| *S1 scenario* | | | | | |
| SCC PRICES/tC | –744/year | | –365/year | | |
| US$5[1] | | –74,312 | | –70,650 | –144,962 |
| US$50[2] | | –743,120 | | –706,496 | –1,449,616 |
| US$312[3] | | –4,637,069 | | –4,408,533 | –9,045,602 |
| DECC PRICES/$tCO_2$ | –744/year | –7,857,364 | –365/year | –7,396,792 | –15,254,156 |
| *S2 scenario* | | | | | |
| SCC PRICES/tC | –744/year for 20 years | | –365/year | | |
| | –947/year for the following 30 years | | | | |
| US$5[1] | | –82,410 | | –70,650 | –153,060 |
| US$50[2] | | –842,099 | | –706,496 | –1,548,595 |
| US$312[3] | | –5,142,380 | | –4,408,533 | –9,550,913 |
| DECC PRICES/$tCO_2$ | –744/year for 20 years | –9,109,158 | –365/year | –7,396,792 | –16,505,950 |
| | –947/year for the following 30 years | | | | |
| *S3 scenario* | | | | | |
| SCC PRICES/tC | –744/year for 20 years | | –45,000/year | | |

(continued)

**Table 10.2** (continued)

| Scenario | Saltmarshes | | Seagrasses | | European blue carbon ecosystems |
|---|---|---|---|---|---|
| | Area lost (ha) | Carbon storage service loss economic value by 2060 (US$) | Area lost (ha) | Carbon storage service loss economic value by 2060 (US$) | Total carbon storage service loss economic value by 2060 (US$) |
| US$5[1] | −947/year for the following 30 years | −82,410 | | −8,708,327 | −8,790,737 |
| US$50[2] | | −842,099 | | −87,083,265 | −87,925,364 |
| US$312[3] | | −5,142,380 | | −543,399,576 | −548,541,956 |
| DECC PRICES/$tCO_2$ | −744/year for 20 years | −9,109,158 | −45,000/year | −911,972,861 | −921,082,019 |
| | −947/year for the following 30 years | | | | |

[1]Pearce 2003
[2]Stern 2007
[3]Tol 2005

## 10.5 Conclusions

As reported by Luisetti et al. (2013, 2014), given the current rate of loss of these ecosystems and the global financial crises of the last decade, there is a risk that future generations will be worse-off economically, in terms of ecosystem services availability than contemporary society. The second case study highlighted that the carbon storage potential in *P. oceanica* is the greatest, but also that this habitat has the higher risk of depletion. The uncertainty surrounding threshold limits in terms of their spatial and temporal scale calls for the adoption of a precautionary approach to maintain coastal blue carbon ecosystems to ensure a minimum of functionality. This chapter aims to raise the profile of blue carbon to maintain this climate regulation service through the conservation of the ecosystems providing it. The restoration case for both saltmarsh and seagrass is further strengthened when the co-benefits they provide are considered (e.g. saltmarshes provide recreation and amenity benefits, fish nursery benefits and flood protection enhancement; and seagrasses are known to provide similar services). Limitations in natural and social sciences data for site-specific ecosystem services valuation may limit the economic analysis and its role in policy and project appraisal. There does, however, seem to be some scope for judicious use of the value (benefit) transfer method to provide missing primary valuation data at other sites of interest. However, although the first case study demonstrates that at the UK level, at least for the carbon storage service, biophysical and economic estimates can be transferred without any adjustment, Luisetti et al. (2014) warn against an extensive use of this practice when bundles of ecosystem services are involved.

## References

Adams, C. A., Andrews, J. E., & Jickells, T. D. (2012). Nitrous oxide and methane fluxes vs. carbon and nutrient burial in new intertidal and salt marsh sediments. *Science of the Total Environment, 434*, 240–251.

Allen, J. R. L., & Rae, J. E. (1987). Late Flandrian shoreline oscillations in The Severn Estuary: A geomorphological and stratigraphical reconnaissance. *Philosophical Transactions of the Royal Society of London, Series B, 315*, 185–230.

Andrews, J. E., Samways, G., Dennis, P. F., & Maher, B. A. (2000). Origin, abundance and storage of organic carbon and sulphur in the Holocene Humber Estuary - Emphasising human impact on storage changes. In I. Shennan & J. E. Andrews (Eds.), *Holocene land-ocean interaction and environmental change around the North Sea* (Geological Society special publication No. 166, pp. 145–170). London: Geological Society.

Andrews, J. E., Burgess, D., Cave, R. R., Coomes, E. G., Jickells, T. D., Parkes, D. J., & Turner, R. K. (2006). Biogeochemical value of managed realignment, Humber estuary UK. *Science of the Total Environment, 371*, 19–30.

Andrews, J. E., Samways, G., & Shimmield, G. B. (2008). Historical storage budgets of organic carbon, nutrient and contaminant elements in saltmarsh sediments: Biogeochemical context for managed realignment, Humber Estuary, UK. *Science of the Total Environment, 405*, 1–13.

Andrews, J. E., Jickells, T. D., Adams, C. A., Parkes, D. J., & Kelly, S. D. (2011). Sediment record and storage of organic carbon and the nutrient elements (N, P, and Si) in estuaries and near-coastal seas. In E. Wolanskiand & D. S. McLusky (Eds.), *Treatise on estuarine and coastal science* (Vol. 4, pp. 9–38). Waltham: Academic.

Banzhaf, H. S., & Boyd, J. (2012). The architecture and measurement of an ecosystem services index. *Sustainability, 4*, 430–461.

Barbier, E. B., Hacker, S. D., Kennedy, C., Koch, E. W., Stier, A. C., & Silliman, B. R. (2011). The value of estuarine and coastal ecosystem services. *Ecological Monographs, 81*, 169–193.

Beaumont, N. J., Jones, L., Garbutt, A., Hansom, J. D., & Toberman, M. (2014). The value of carbon sequestration and storage in coastal habitats. *Estuarine, Coastal & Shelf Science, 137*, 32–40.

Bianchi, T. S. (2007). *Biogeochemistry of estuaries*. Oxford: Oxford University Press.

Boyes, S. J., & Elliott, M. (2006). Organic matter and nutrient inputs to the Humber Estuary, England. *Marine Pollution Bulletin, 53*(1-4), 136–143.

Cebrian, J., Duarte, C., Marba, N., & Enriquez, S. (1997). Magnitude and fate of the production of four co occurring western Mediterranean seagrass species. *Marine Ecology Progress Series, 155*, 29–44.

Chevallier, J. (2010). *Carbon prices during the EU ETS phase II: Dynamics and volume analysis*. http://econpapers.repec.org/paper/halwpaper/halshs-00459140.htm

Chmura, G. L., Anisfeld, S. C., Cahoon, D. R., & Lynch, J. C. (2003). Global carbon sequestration in tidal, saline wetland soils. *Global Biogeochemical Cycles, 17*, 1111.

Costanza, R., d'Arge, R., de Groot, R. S., Farber, S., Grasso, M., Hannon, B., Limburg, K., Naeem, S., O'Neil, R. V., Paruelo, J., Raskin, R. G., Sutton, P., & van den Belt, M. (1997). The value of the world's ecosystem services and natural capital. *Nature, 387*, 253–260.

Costanza, R., de Groot, R., Sutton, P., van der Ploeg, S., Anderson, S. J., Kubiszewski, I., Farber, S., & Turner, R. K. (2014). Changes in the global value of ecosystem services. *Global Environmental Change, 26*, 152–158.

da Silva Copertino, M. (2011). Add coastal vegetation to the climate critical list. *Nature, 473*(7347), 255.
Dagley, J. R. (1995). *Northey Island managed retreat scheme* (English Nature Research Reports, No. 128). Peterborough: English Nature.
Department of Energy and Climate Change (DECC). (2011). *A brief guide to the carbon valuation methodology for UK policy appraisal*. London: Department of Energy and Climate Change.
Department of Energy and Climate Change (DECC). (2013). Updated short-term traded carbon values used for UK public policy appraisal. 16 Sept 2013.
Edwards, A. M. C., & Winn, P. S. J. (2006). The Humber Estuary, Eastern England: Strategic planning of flood defences and habitats. *Marine Pollution Bulletin, 53*, 165–174.
Elliott, M., Cutts, N. D., & Trono, A. (2014). A typology of marine and estuarine hazards and risks as vectors of change: A review for vulnerable coasts and their management. *Ocean & Coastal Management, 93*, 88–99.
Fisher, B., Turner, R. K., & Morling, P. (2009). Defining and classifying ecosystem services for decision making. *Ecological Economics, 68*(3), 643–653.
Fourqurean, J.W. et al. (2012). Seagrass ecosystems as a globally significant carbon stock. *Nature Geoscience. 5*, 505–509. doi:http://www.nature.com/ngeo/journal/v5/n7/abs/ngeo1477.html#supplementary-information.
Gacia, E., Duarte, C. M., & Middelburg, J. J. (2002). Carbon and nutrient deposition in a Mediterranean seagrass (*Posidonia oceanica*) meadow. *Limnology and Oceanography, 47*(1), 23–32.
Gómez-Baggethun, E., De Groot, R., & Lomas, P. L. (2010). The history of ecosystem services in economic theory and practice: From early notions to markets and payment schemes. *Ecological Economics, 69*(6), 1209–1218.
Heal, G. (2000). *Nature and the marketplace*. Washington, DC: Island Press.
HMT. (2011). *Green Book: Appraisal and evaluation in Central Government*. London: HMSO.
Howarth, R. B., & Farber, S. (2002). Accounting for the value of ecosystem services. *Ecological Economics, 41*, 421–429.
Jickells, T. D., & Weston, K. (2011). Nitrogen cycle – External cycling: Losses and gains. In E. Wolanskiand & D. S. McLusky (Eds.), *Treatise on estuarine and coastal science* (Vol. 5, pp. 261–278). Waltham: Academic.
Jickells, T., Andrews, J., Samways, G., Sanders, R., Malcolm, S., Sivyer, D., Parker, R., Nedwell, D., Trimmer, M., & Ridgway, J. (2000). Nutrient fluxes through the Humber estuary – Past, present and future. *Ambio, 29*, 130–135.
Jones, L. et al. (2011). Chapter 11: Coastal margins. In: *The UK National Ecosystem Assessment technical report*. UK National Ecosystem Assessment, UNEP-WCMC, Cambridge.
Jorda, G., Marba, N., & Duarte, C. M. (2012). Mediterranean seagrass vulnerable to regional climate warming. *Nature Climate Change, 2*(11), 821–824.
Kennedy, H., & Bjork, M. (2009). Seagrass meadows. In D. Laffoley & G. D. Grimsditch (Eds.), *The management of natural coastal carbon sinks* (pp. 23–28). Gland: IUCN.
Langmead, O., McQuatters-Gollop, A., & Mee, L. D. (2007). *European lifestyles and marine ecosystems: Exploring challenges for managing Europe's seas* (Vol. 43). Plymouth: University of Plymouth Marine Institute.
Luisetti, T., Turner, R. K., Bateman, I. J., Morse-Jones, S., Adams, C., & Fonseca, L. (2011). Coastal and marine ecosystem services valuation for policy and management: Managed realignment case studies in England. *Ocean & Coastal Management, 54*(3), 212–224.
Luisetti, T., Jackson, E. L., & Turner, R. K. (2013). Valuing the European "coastal blue carbon" storage benefit. *Marine Pollution Bulletin, 71*, 101–106.
Luisetti, T., Turner, R. K., Jickells, T., Andrews, J., Elliott, M., Schaafsma, M., Beaumont, N., Malcolm, S., Burdon, D., Adams, C., & Watts, W. (2014). Coastal Zone Ecosystem Services: From science to values and decision making; a case study. *Science of the Total Environment, 493*, 682–693.

Matsui, N., Morimune, K., Meepol, W., & Chukwamdee, J. (2012). Ten year evaluation of carbon stock in mangrove plantation reforested from an abandoned shrimp pond. *Forest, 3*(4), 431–444.

Mazik, K., Musk, W., Dawes, O., Solyanko, K., Brown, S., Mander, L., & Elliott, M. (2010). Managed realignment as compensation for the loss of intertidal mudflat: A short term solution to a long term problem? *Estuarine, Coastal and Shelf Science, 90*, 11–20.

Mcleod, E., Chmura, G. L., Bouillon, S., Salm, R., Björk, M., Duarte, C. M., Lovelock, C. E., Schlesinger, W. H., & Silliman, B. R. (2011). A blueprint for blue carbon: Toward an improved understanding of the role of vegetated coastal habitats in sequestering $CO_2$. *Frontiers in Ecology and the Environment, 9*(10), 552–560.

McLusky, D. S., & Elliott, M. (2004). *The estuarine ecosystem: Ecology, threats and management* (3rd ed.). Oxford: Oxford University Press.

Middelburg, J. J., & Levin, L. A. (2009). Coastal hypoxia and sediment biogeochemistry. *Biogeosciences, 6*, 1273–1293.

Morgan, P. A., & Short, F. T. (2002). Using functional trajectories to track constructed salt marsh development in the Great Bay estuary, Maine, New Hampshire, U.S.A. *Restoration Ecology, 10*, 461–473.

Mossman, H. L., Brown, M. J. H., Davy, A. J., & Grant, A. (2012). Constraints on salt marsh development following managed coastal realignment: Dispersal limitation or environmental tolerance? *Restoration Ecology, 20*, 65–75.

Murray, B. C., Pendleton, L., Jenkins, W. A., & Sifleet, S. (2011). *Green payments for blue carbon: Economic incentives for protecting threatened coastal habitats*. Durham: Nicholas Institute for Environmental Policy Solutions, Duke University.

Nellemann, C., Corcoran, E., Duarte, C.M., Valdés, L., De Young, C., Fonseca, L., Grimsditch, G. (Eds). (2009). *Blue carbon*. A rapid response assessment. Norway: United Nations Environment Programme, GRID-Arendal. Birkeland Trykkeri AS.

Pearce, D. (2003). The social cost of carbon and its policy implications. *Oxford Review of Economic Policy, 19*(3), 362.

Pendleton, L., Donato, D. C., Murray, B. C., et al. (2012). Estimating global "Blue Carbon" emissions from conversion and degradation of vegetated coastal ecosystems. *PLOS One, 7*(9), e43542.

Rees, J. G., Ridgway, J., Ellis, S., Knox, R. W. O., Newsham, R., & Parkes, A. (2000). Holocene sediment storage in the Humber Estuary. In I. Shennan & J. E. Andrews (Eds.), *Holocene land-ocean interaction and environmental change around the North Sea* (Geological Society special publication No. 166, pp. 119–143). London: Geological Society.

Rickels, W., Görlich, D., & Oberst, G. (2010). *Explaining European emission allowance price dynamics: Evidence from phase II* (Kiel working papers No. 1650). Kiel: Kiel Institute for World Economy.

Seitzinger, S., Harrison, J. A., Bohlke, J. K., Bouwman, A. F., Lowrance, R., Peterson, B., Tobias, C., & Van Drecht, G. (2006). Denitrification across landscapes and waterscapes: A synthesis. *Ecological Applications, 16*, 2064–2090.

Shepherd, D., Burgess, D., Jickells, T., Andrews, J., Cave, R., Turner, R. K., Aldridge, J., Parker, E. R., & Young, E. (2007). Modelling the effects and economics of managed realignment on the cycling and storage of nutrients, carbon and sediments in the Blackwater estuary UK. *Estuarine and Coastal Shelf Science, 73*, 355–367.

Sifleet, S., Pendleton, L., & Murray, B. C. (2011). *State of the science on coastal blue carbon: A summary for policy makers. Nicholas Institute for Environmental Policy Solutions*. Durham: Duke University.

Sousa, A. I., Lillebø, A. I., Pardal, M. A., & Caçador, I. (2010). Productivity and nutrient cycling in salt marshes: Contribution to ecosystem health. *Estuarine, Coastal and Shelf Science, 87*(4), 640–646.

Stern, N. (2007). *The economics of climate change: The stern review*. Cambridge, UK: Cambridge University Press.

Thomas, S. (2014). Blue carbon: Knowledge gaps, critical issues, and novel approaches. *Ecological Economics, 107*, 22–38.

Tol, R. S. J. (2005). The marginal damage costs of carbon dioxide emissions: An assessment of the uncertainties. *Energy Policy, 33*, 2064–2074.

Turner, R. K., Paavola, J., Cooper, P., Farber, S., Jessamy, V., & Georgiu, S. (2003). Valuing Nature: Lessons learned and future research directions. *Ecological Economics, 46*, 493–510.

Turner, R. K., Burgess, D., Hadley, D., Coombes, E. G., & Jackson, N. (2007). A cost-benefit appraisal of coastal managed realignment policy. *Global Environmental Change, 17*, 397–407.

Turner, R. K., Georgiou, S., & Fisher, B. (2008). *Valuing ecosystem services: The case of multi-functional wetlands*. London/Washington, DC: Earthscan.

UK National Ecosystem Assessment Follow-on. (2014). *The UK National Ecosystem Assessment Follow-on: Synthesis of the key findings*. Cambridge: UNEP-WCMC, LWEC.

Ullman, R., Bilbao-Bastida, V., & Grimsditch, G. (2013). Including blue carbon in climate market mechanisms. *Ocean & Coastal Management, 83*, 15–18.

Vandergeest, P., Flaherty, M., & Miller, P. (2009). A political ecology of shrimp aquaculture in Thailand. *Rural Sociology, 64*(4), 573–596.

Wolters, M., Garbutt, A., Bekker, R. M., & Carey, P. D. (2008). Restoration of salt-marsh vegetation in relation to site suitability, species pool and dispersal traits. *Journal of Applied Ecology, 45*, 904–912.

# Chapter 11
# What Future for the English Coastline? Managed Realignment Benefits and Their Value Estimate Transferability

**Tiziana Luisetti, R.K. Turner, Tim Jickells, Julian Andrews, Michael Elliott, Marije Schaafsma, Nicola Beaumont, Stephen Malcolm, Daryl Burdon, Christopher Adams, and William Watts**

## 11.1 Introduction

One of the two case studies (Luisetti et al. 2014) reviewed in Chap. 10 was concerned with the economic valuation and estimate transferability of the stock and flows of the carbon storage service provided by the saltmarsh habitats in two estuaries located along the east coast of England. Intertidal vegetated ecosystems of the Humber and the Blackwater estuaries also provide other services (Turner et al.

T. Luisetti (✉) • S. Malcolm
Centre for Environment, Fisheries & Aquaculture Science (CEFAS), Pakefield Road, Lowestoft, Suffolk NR33 0HT, UK
e-mail: tiziana.luisetti@cefas.co.uk

R.K. Turner
School of Environmental Sciences, University of East Anglia, Norwich, UK

T. Jickells
Centre for Ocean and Atmospheric Sciences (COAS), School of Environmental Sciences, University of East Anglia, Norwich, UK

J. Andrews • C. Adams
School of Environmental Sciences, University of East Anglia, Norwich, Norfolk NR4 7TJ, UK

M. Elliott • D. Burdon
Institute of Estuarine & Coastal Studies, University of Hull, Hull HU6 7RX, UK

M. Schaafsma
Geography and Environment and Centre for Biological Sciences, University of Southampton, University Road, Southampton SO17 1BJ, UK

N. Beaumont
Plymouth Marine Laboratory, Prospect Place, Plymouth, Devon PL1 3DH, England

W. Watts
Environment Agency, 7 Canon Beck Road, London SE 16 7DJ, UK

 
R.K. Turner, M. Schaafsma (eds.), *Coastal Zones Ecosystem Services*, Studies in Ecological Economics 9, DOI 10.1007/978-3-319-17214-9_11

2007; Luisetti et al. 2011a), for example: fish nurseries which lead to food production; and amenity and recreational benefits (e.g. walking and enjoying the scenery of the coast, bird-watching, sea angling etc).

As reported in Luisetti et al. (2014) and in Turner & Luisetti (2015), in the last few centuries coastal areas have been under local pressures such as land conversion and reclamation, agricultural regime changes and nutrient run-off, and general pollution. Global pressures include global warming direct effects such as sea level rise, sometimes resulting in intertidal habitats being 'squeezed' between the sea and sea defence or coastal protection structures (Doody 2004); and indirect effects such as storm surges and inundation.

Despite the scientific uncertainty surrounding climate change impacts especially at the regional/local scale (e.g. input data in climate models; the methodologies used to produce climate models and the limitations of the UK Climate Projections (UKCP)) enough evidence has been accumulated to justify further mitigation and adaptation policy measures.

The UKCP09[1] (2009) projections for UK coastal absolute sea level rise (not including land movement) for the period 1990–2095 are in the range 12–76 cm. Projected extreme surge level is not expected to increase by more than 9 cm by 2100. Excluding sea level rise, over the twenty-first century the size of surge is expected to occur on average about once in 50 years increasing by less than 0.9 mm per annum. Nevertheless, since the dramatic storm surge of 1953, the North Sea has recently suffered two major storm surges – one in 2007 and one in 2013.

Actions to tackle the risk of flooding in coastal areas depend on the area at risk (e.g. adaptation measures to climate change in The Netherlands will be different from those in the UK). In the UK, there are several on-going experimental projects. The Thames Estuary 2100 project is an attempt to move from a reactive flood defence to a proactive flood risk management focused on the barrier to protect London and its estuary from tidal surges entering from the North Sea. Other experiments relate to managed realignment (MR) of the coastal zone. MR schemes (Elliot et al. 2014; Mazik et al. 2010) are measures that may be considered to be so called "low regret" policies as they are in some contexts a more sustainable coastal defence option compared to the traditional hard defences (e.g. sea walls). In practice MR schemes have involved the voluntary breaching of sea walls that have almost reached their expected life, letting the sea flood managed areas (usually agricultural land) behind the sea wall. Because some coastal land is sacrificed, MR is considered controversial (Royal Commission 2010). Mixed approaches in which only high-value areas continue to be protected and the rest of the coastline is left free to adapt to change more naturally seems to be a pragmatic long-term solution (Turner et al. 2007). However, the reality of coastal management and politics is complex with national strategic decision making sometimes at odds with more regional/local interests.

Previous studies (Turner et al. 2007; Luisetti et al. 2011a) showed the economic efficiency of MR schemes in these estuaries, taking into account trade-offs such as

---

[1] UKCP09 stands for UK Climate Projections. UKCP09 is founded by the UK Department for Environment, Food and Rural affairs (DEFRA). It is based on sound science and projections provided by the Met Office Hadley Centre. It updates the 2002 results of the UKCIP02.

lost agricultural land, and depending on the chosen socio-economic future scenario, the discount rate and time horizon considered. In Luisetti et al. (2011a) the economic efficiency advantages of MR schemes were apparent under any of the three scenarios considered in the Blackwater estuary case study over a time horizon of 50/100 years. However, the authors also found an anomaly (inefficient outcome) with the scenario that included the highest level of realignment/sustainability and when a short time period such as 25 years was taken into account. The anomaly could be explained by the higher weight in the discounted valuation of the realignment costs (breaching existing sea walls and building secondary lines of defence) that are usually incurred in the first year of a project, compared to the weight of the subsequent benefits considered for the same short period of time (e.g. 25 years) even though these benefits actually accrue over a longer period of time.

In this chapter, after reviewing current British coastal governance issues, we summarise the results of the rest of the case study presented in Chap. 10 and described in detail in Luisetti et al. (2014). The reminder of the chapter is concerned with the investigation of the transferability of biophysical and economic value estimates of the fish nursery and recreation and amenity services between estuaries. This is of particular importance as ecosystem services are context and scale dependent (Fisher et al. 2009), and valuation practices and results should reflect this. Value transfer is a method by which the value estimated for an ecosystem service or environmental benefit in a specific area (the study site) and time is applied to another area (the policy site) and time (Navrud and Ready 2007).

In the study by Luisetti et al. (2014) the limited availability of biophysical data for the fish nursery service was noted, but it still presents a useful analysis of the issues that should be considered when the transferability of willingness to pay (WTP) estimates are concerned. The study investigated the issues surrounding the transfer of WTP values previously estimated in Luisetti et al. (2011a) for the Blackwater to the Humber estuary. Their results throw light on the appropriate use of the value transfer technique when ecosystem services are concerned and highlight the need for a dialogue between different disciplines encompassing natural (e.g. biogeochemistry, ecology, marine biology) and social sciences, as well as environmental economics for improved decision making. This chapter offers some suggestions when data availability limitations are encountered, and present for each case study a set of assumptions meeting the minimum conditions required for value transfer.

An important part of the contextual background in any valuation and/or benefits transfer exercise is the prevailing system of governance and we look at this issue next.

## 11.2 British Coastal Policy and Governance Structure: The Case of Managed Realignment

Turner & Luisetti (2015) present a detailed review of the coastal policy and governance structure in England, which we summarise and update here. Population and assets located in British coastal communities, especially in the east and south of England, are potentially at risk because of natural threats. Increased cliff erosion and more frequent flooding is speeding up the degradation of existing defences. This is putting at risk private assets and several ecosystem services reducing coastal societal well-being. At times of limited financial budgets, coastal issues highlight the different needs at the national strategic level and the local level, which may require either coastal protection or compensation (O'Riordan et al. 2008).

Nicholson-Cole and O'Riordan (2009) stress the need for practical responses to the difficulties posed by coastline change to human welfare. Prospects for change may be provided by the adoption of a more adaptive management and coastal governance. However, Nicholson-Cole and O'Riordan highlight that the actual response process has been fragmented and there has been a lack of plan and policy integration among public, private and voluntary organisations interests concerned with coastal adaptation.

In 2002 the European Commission encouraged Member States to implement an integrated coastal zone management (ICZM) (2002/413/EC), which is slowly becoming embedded in national strategies. In 2005 the British Department for Environment, Food and Rural Affairs (Defra) launched the Making Space for Water (MSFW) programme (Defra 2005). MSFW represents the new policy structure for British coastal management; a policy that shifts from the hard defences to more sustainable forms of flood protection (e.g. including among other measures managed realignment) recognising that risk can be reduced but not eliminated (POST 2009).

O'Riordan et al. (2008) reviewed the governance of coastal areas for England and Wales that, up to 2009, had developed over three levels: political (central government departments), executive (statutory but non-departmental agencies) and civic organisations and coastal communities (other organisations concerned with the coastal management). The first level contains: Defra which provides the overall strategic policy as well as the supervision of the non-departmental agencies; the Department of Energy and Climate Change (DECC), which is concerned with climate change mitigation polices (previously on the remit of Defra); the Department for Communities and Local Government (DCGL) that produces planning policy and local government finance; and the Treasury, which provides general funding to Defra, the non-departmental agencies and local authorities. Local Government County and District Councils operate both at the political and at the executive level. At the second level we find the two relevant governmental agencies, Environment Agency (EA) and Natural England (NE), and the Local Maritime Authority. A governance problem existed here as these two agencies have different objectives and statutory duties but both have an interest in planning the management of the coastline. At the third level, there are local residents and communities, and non-governmental bodies, which in some cases also own coastal land managed as natural reserves.

In 2009, the UK Government passed the Marine and Coastal Access Act (2009) and approved the creation of the Marine Management Organisation (MMO), which offered an opportunity to harmonise management. But Boyes and Elliot (2015) however maintain that many overlapping responsibilities still exist, with the MMO acting as the regulator for most but not all of the marine environment and related economic activities. So despite the 2009 Act, the well levelled governance structure described in O'Riordan et al. (2008) still faces coordination challenges. The three governance levels are interconnected so that both co-ordination and organisational problems often arise, which in turn lead local people to question policy and financial decision-making in coastal areas sometimes taking their own actions (for example by constructing their own ad hoc coastal defences). Nevertheless, Boyes and Elliott (2015) stress the importance of the Marine and Coastal Access Act, as a piece of legislation that at least points in the direction of an ecosystem-based approach for the integrated management of marine and coastal areas.

An adaptive management approach (see Chaps. 1 and 2) would also stress equity as well as efficiency in coastal management. O'Riordan et al. (2008), for example, recommend much more clarity over discretionary public defence of any person or property on the coast, together with the legal possibility for private owners to provide their own coastal protection as a matter of social justice. They also suggest that coastal policy in England should become much more long term and strategic with new approaches to land-use planning that take into account enhanced coastal flood and erosion risk for its implementation.

## 11.3 Transferring Benefit Values Within the Same Region of the Same Country: A Case Study Located on the East Coast of England

Ecosystem services values tend to be spatially explicit and context dependency may limit the scope for the aggregation of values at the spatial scale. On the other hand, as fieldwork and local data collection across numerous individual sites can be very expensive, value transfer techniques are often used instead to obtain approximate data for biophysical and economic value estimates in non-surveyed area of interest (known in the literature as the 'policy site').

In their original MR studies Turner et al. (2007) and Luisetti et al. (2011a) analysed the efficiency of MR in the Humber and the Blackwater respectively, allowing for some sacrifice of lower quality agricultural land. However, areas in which the risk of loss or damage may raise specific significant ethical concerns (e.g. urban areas including a range of assets and private properties) were deliberately avoided. In Luisetti et al. (2014) the possibility to transfer biophysical estimates and welfare value estimates for the ecosystem services provided by the Blackwater estuary to the Humber estuary despite the limiting characteristic of context dependency is investigated. Chapter 10 reviewed the case study for the single service of carbon sequestration and storage and its value estimate transfer. In the case of carbon storage, current data suggested similar contexts between the Humber and the Blackwater areas, and hence their almost straightforward transferability. In the case studies reviewed in this chapter too it is assumed that since both estuaries are located on the same coastline they are geo-morphologically and socio-economically similar (see Chap. 10 for a description of the case study area), but the values being transferred cover a number of ecosystem services.

### 11.3.1 Carbon

In Chap. 10 we reported that transferability of biophysical estimates appears to be broadly valid for regional carbon storage (Luisetti et al. 2014). Furthermore, at the European scale Luisetti et al. (2013) demonstrated that transferability is possible but that it is more challenging and requires a larger number of assumptions – from the carbon burial data estimates to the mapping. We can infer that at higher scales (e.g. global scale) the number of assumptions will increase considerably pushing up the level of uncertainty over the biophysical estimates and ultimately over the economic values attached to them.

### 11.3.2 Fisheries

Luisetti et al. (2014) argue that food (fish, via fisheries production in nursery grounds) production transferability may also be possible. However, they considered context dependency a major issue for fisheries productivity in saltmarshes because ecological conditions and functions may differ significantly between estuaries. They report that production estimates within managed realignment sites in the Humber estuary at the time of the analysis were less-advanced than for the Blackwater estuary (Burdon et al. 2011; Fonseca 2009; Luisetti et al. 2011a). Hence the data could not be reliably transferred without a more in depth analysis and possibly a targeted data collection effort (e.g. fish community analysis, age of population, survival rate, etc). The authors, however, advocate investigation of transferability options for the Humber. Future research, they suggest, may include the appropriateness of using the same production function estimated for the Blackwater and applying the data collected for the Humber to this production function, but also the estimation of a new specific production function for the Humber using Humber data and a validity exercise comparing those results with the estimates transferred from the Blackwater estuary to the Humber.

### 11.3.3 Recreation

In Luisetti et al. (2014) the authors revised the results of Luisetti et al. (2011a) in order to transfer the results of the willingness to pay (WTP) values for re-created saltmarshes in the Blackwater estuary obtained by a choice experiment survey to the Humber. The attributes used in the choice experiment for the Blackwater were: the extent of the saltmarsh area re-created (AREA); the number of protected bird species observable (BIRD); the distance from respondent's home to the new recreational site (DISTANCE); the potential public access to the new area (ACCESS); and the hypothetical annual cost for the respondents expressed by an increment in their local council tax to implement MR schemes and re-create saltmarshes (TAX). The attribute *ACCESS* reveals how much people value the possibility of 'using' the newly created intertidal habitat. The variable *BIRD* was used as a proxy for non-use (e.g. biodiversity) value.

Two futures scenarios were considered: a future scenario (F1) in which saltmarshes are re-created through MR; and a future scenario (F2) in which, mostly because of climate change effects, salt marsh areas continue to be lost. In Luisetti et al. (2014), three main issues were critically analysed. The first relates to the WTP values obtained for saltmarsh re-creation. In the original study (Luisetti et al. 2011a) the choice experiment conducted in the Blackwater addressed saltmarsh creation. In the Luisetti et al. (2014) study environmental losses are also valued using WTP whereas it could be argued that they should have been more appropriately valued through the so called Willingness-to-Accept (WTA) compensation for a loss measure. This is because it has been observed (Pearce 2002) that, primarily because of substitution effects and loss aversion (Kahneman and Tversky 1984), WTA for losses seem to be larger than WTP for gains. However, it has also been observed that when there are sufficient substitution options for the good lost, the divergence between WTP and WTA decreases (Hanemann 1999). The substitute site situation was investigated in the study by Luisetti et al. (2011a). The survey conducted for the choice experiment included a test in which two maps with different scales, presenting the substitutes in the area, were shown to the respondents. Since the test revealed that respondents were aware of the many substitutes close to the Blackwater, Luisetti et al. (2014) assumed that WTP values for gains could be used as a proxy for WTA values for losses. The estimates are probably conservative because in Luisetti et al. (2011a) the WTP function was estimated for areas between 10 and 70 ha, but in the case study of Luisetti et al. (2014) the two scenarios considered area changes greater than 70 ha. Furthermore, in a parallel study, Luisetti et al. (2011b) report an anomaly related to the range of levels used for the attribute *distance* used in the choice experiment. Taking account of these problems, the aggregated WTP values in Luisetti et al. (2014) do not account for potential positive WTP/WTA of people beyond 32 miles from the newly created saltmarshes.

A further issue is related to the additive characteristic of the original WTP function: i.e. total WTP is a summation of the monetary values of the attributes AREA, DISTANCE, ACCESS and BIRD. This means that WTP values for distance, access and bird species are not scaled to the extent of the saltmarsh area. So Luisetti et al. (2014) highlight that even for very small increases in saltmarsh area this function will produce relatively high WTP estimates, whilst the added hectares may not actually be able to support bird species or be of recreational interest. The authors also consider that the effect of area changes is included in the WTP function after applying a log transformation, which generally implies that the additional WTP for an additional hectare decreases as the total area increases, reflecting theoretically expected satiation effects. However, the authors note that this also suggests that for very small changes (e.g., <5 ha) the WTP per ha for the attribute AREA is unrealistically high, whilst for additional saltmarshes of much more than 70 ha (the upper boundary of the variable), the WTP value is not much higher than for saltmarshes of

70 ha. To address the second and third issue, the Luisetti et al. (2014) study assumed that the function was applicable to areas <10 ha. For the large gains in area, as well as for losses of >70 ha, the authors provided two estimates: option A, which is based on the assumption that the WTP functions can be used beyond the range of the AREA variable without adjustments; and option B, which uses the additional value per household for an increase from 69 to 70 ha new saltmarsh to any additional ha of saltmarsh beyond >70 ha, assuming a linear function from that point onwards. In the case study in Luisetti et al. (2014), calculations for the estimates of ecosystem services provided by the estuaries area are also based on the extent of saltmarshes estimated for each scenario in previous studies (Luisetti et al. 2011a; Turner et al. 2007; Andrews et al. 2000).

The resulting WTP function from the study by Luisetti et al. (2011a) was used in Luisetti et al. (2014). Since significant distance decay values were revealed in the econometric analysis of the CE in Luisetti et al. (2011a), the authors took this into account in the aggregation of WTP estimates per households over the relevant population. For a detailed explanation of the methodology used to calculate the aggregate WTP values (i.e. recreational benefit flows) see Luisetti et al. (2014).

The estimated present values (PV) of the recreational benefit flows over a time horizon of 100 years were discounted following the declining discount rate scheme recommended in the UK Green Book (HMT 2011). As explained in Chap. 10, for the (F1) scenario, the PV are estimated based on the assumption that benefits only start to accrue after 5 years in the Blackwater estuary MR site. As reported in Luisetti et al. (2014) values of this scenario range from GBP 78 million (use values only, option A) to GBP 328 million (use and non-use values, option B). The (negative) benefit flows of the (F2) scenario increase each year as more saltmarsh is lost and vary between GBP 43 million (option A, use values) to GBP 86 million (option B, use and non-use values). Option B results in considerably higher benefit flows than option A for the (F1) scenario, whilst for the (F2) scenario the differences are fairly small, partly because the total saltmarsh loss in 2110 is 244 ha.

Finally, the economic value of the amenity and recreation service estimated for the Blackwater estuary was transferred to the Humber estuary making use of a benefit transfer value technique. The authors argue that using a UK based study is likely to provide reasonably accurate estimates because the socio-economic conditions and ecosystem services provided are likely to be more similar than in a global meta-analysis. For examples, the authors find that although the Humber estuary surroundings are more populated than those of the Blackwater, socio-economic characteristics appear to be similar (Office National Statistics 2011). Therefore, the same coefficients for the WTP function in Luisetti et al. (2011a) were used. Table 11.1 summarises the results.

As explained in Chap. 10, in the Humber study, the recreation and amenity benefits in the (F1) scenario are not assumed to start until 15 years after the establishment of the MR site (176 ha). In the (F2) scenario, the total area lost in 2100 is 2,500 ha, immediately leading to losses in recreation and amenity values. The estimated benefits of the (F1) scenario range from GBP 33 million (option A, use values) to GBP 48 million (option B, use and non-use values). The difference between the estimated benefits of the (F1) scenario and the losses under the (F2) scenario are

**Table 11.1** Present value of the aggregated willingness to pay (WTP) flows for recreation in the Humber saltmarshes under the different scenarios and over a time horizon of 100 years (2010–2110) discounted following the declining discount rate scheme of the UK Green Book (HMT 2011) (£ *1000). Source: Luisetti et al. (2014)

| Humber | | Option A | Option B |
|---|---|---|---|
| Fl scenario | *WTP use and non-use values* | 45,901 | 48,268 |
| | *WTP mainly use values* | 33,187 | 35,554 |
| F2 scenario | *WTP use and non-use values* | −76,874 | −126,611 |
| | *WTP mainly use values* | −57,281 | −106,281 |

Note: In option A, the WTP function is used beyond the range of the *AREA* variable without adjustments. Option (B) applies the additional value per household for an increase from 69 to 70 ha new saltmarsh (e.g. £0.0156/year/ha for any additional ha of saltmarsh beyond >70 ha)

larger using the calculation method of option B than under option A. This is because of the diminishing marginal value assumed for any additional ha under option A, which implies that the much larger area lost in the (F2) scenario (compared to (F1)) does not result in a proportional decrease in monetary value.

## 11.4 Conclusion

The results of the case studies examined here for the recreation and amenity service, the fish nursery service, and the carbon storage service (Chap. 10), suggest that because of socio-ecological system complexity the available biophysical and socio-economic value estimates and their suitability for benefit transfer should be critically assessed before undertaking any benefit transfer exercise. The review of these case studies highlights the flexibility that is necessary for an appropriate management of ecosystem services. Because of context dependency and limited data availability, Luisetti et al. (2014) did not transfer the value of the fish nurseries service from the Blackwater to the Humber estuary. By contrast, transfer of biophysical estimates within the same region on the east coast of England for the carbon storage service did not pose any significant transfer issue. Transferability of the recreation and amenity service has been shown to be possible but only subject to several assumptions given the existing data. In all cases, as the scale over which the transfer is attempted increases (up to the national or European level for example) the assumptions made about biophysical estimates and the need to adapt the willingness to pay function of the case study area to the policy site may increase in number and complexity. This clearly mitigates against an extensive use of the value transfer practice for decision making. For the fish nursery case study for example, more location specific data are necessary before proceeding to value transfer. As a consequence, Luisetti et al. (2014) highlight the need of careful consideration in the use and application of value transfer when ecosystem services are concerned both in terms of biophysical data and welfare value estimates if as reliable as is feasible information is to be provided for policy making.

## References

Andrews, J. E., Samways, G., Dennis, P. F., & Maher, B. A. (2000). Origin, abundance and storage of organic carbon and sulphur in the Holocene Humber Estuary – Emphasising human impact on storage changes. In I. Shennan & J. E. Andrews (Eds.), *Holocene land-ocean interaction and environmental change around the North Sea* (Geological society special publication, No. 166, pp. 145–170). London: Geological Society.

Boyes, S. J., & Elliott, M. (2015). The excessive complexity of national marine governance systems – Has this decreased in England since the introduction of the Marine and Coastal Access Act 2009? *Marine Policy, 51*, 57–65.

Burdon, D., Atkins, J. P., & Bhatia, N. (2011). Ecosystem services and societal benefits provided by estuarine saltmarsh (Rep. No. SBB338-F1-2011). Report to Associated British Ports & Environment Agency, Dec 2011.

Defra. (2005). *Making space for water. Taking forward a new government strategy for flood and coastal erosion risk management in England*. London: Defra.

Doody, J. P. (2004). 'Coastal squeeze' – An historical perspective. *Journal of Coastal Conservation, 10*(1–2), 129–138.

Elliott, M., Cutts, N. D., & Trono, A. (2014). A typology of marine and estuarine hazards and risks as vectors of change: A review for vulnerable coasts and their management. *Ocean & Coastal Management, 93*, 88–99.

Fisher, B., Turner, R. K., & Morling, P. (2009). A systems approach to definitions and principles for ecosystem services. *Ecological Economics, 68*, 643–653.

Fonseca, L. (2009). *Fish utilisation of managed realignment areas and saltmarshes in the Blackwater Estuary, Essex, S. E. England*. Ph.D. thesis, Queen Mary University of London

Hanemann, M. (1999). The economic theory of WTP and WTA. In I. Bateman & K. Willis (Eds.), *Valuing environmental preferences: Theory and practice of the contingent valuation method in the US, EU and developing countries* (pp. 42–96). Oxford: Oxford University Press.

HMT. (2011). *Green Book: Appraisal and evaluation in central government*. London: HMSO.

Kahneman, D., & Tversky, A. (1984). Choices, values, and frames. *American Psychologist, 39*(4), 341–350.

Luisetti, T., Turner, R. K., Fonseca, L., & Baker, C. (2011a). Coastal and marine ecosystem service valuation for policy and management. *Ocean and Coastal Management, 54*(3), 212–224.

Luisetti, T., Bateman, I. J., & Turner, R. K. (2011b). Testing the fundamental assumptions of choice experiments: Are values absolute or relative? *Land Economics, 87*(2), 284–296.

Luisetti, T., Jackson, E. L., & Turner, R. K. (2013). Valuing the European "coastal blue carbon" storage benefit. *Marine Pollution Bulletin, 71*, 101–106.

Luisetti, T., Turner, R. K., Jickells, T., Andrews, J., Elliott, M., Schaafsma, M., Beaumont, N., Malcolm, S., Burdon, D., Adams, C., Watts, W. (2014). Coastal zone ecosystem services: From science to values and decision making; a case study. *Science of the Total Environment, 493*, 682–693.

Mazik, K., Musk, W., Dawes, O., Solyanko, K., Brown, S., Mander, L., & Elliott, M. (2010). Managed realignment as compensation for the loss of intertidal mudflat: A short term solution to a long term problem? *Estuarine, Coastal and Shelf Science, 90*, 11–20.

Navrud, S., & Ready, R. (2007). *Environmental value transfer: Issues and methods*. Dordrecht: Springer. 290 pp.

Nicholson-Cole, S., & O'Riordan, T. (2009). Adaptive governance for a changing coastline: Science, policy and the public in search of a sustainable future. In W. N. Adger, I. Lorenzoni, & K. O'Brien (Eds.), *Adapting to climate change: Thresholds, values and governance* (pp. 368–383). Cambridge, UK: Cambridge University Press.

O'Riordan, T., Nicholson-Cole, S., & Milligan, J. (2008). Designing sustainable coastal futures, twenty-first century society. *Journal of the Academy of Social Sciences, 3*, 145–157.

Office for National Statistics, Census 2011.

Pearce, D. (2002). *The role of 'property rights' in determining economic values for environmental costs and benefits*. Report to the Environment Agency. Bristol, Dec 2002.

POST. (2009). *Postnote: Coastal management*, 342. London: Parliamentary Office of Science and Technology.

Royal Commission on Environmental Pollution. (2010). *Adapting institutions to climate change*. London: Crown Copyright.

Turner, R. K., & Luisetti, T. (2015). Toward adaptive management in coastal zones: Experience from the eastern coastline of England. In B. Glavovic, M. Kelly, R. Kay, & A. Travers (Eds.), *Climate change and the coast: Building resilient communities* (pp. 33487–2742). Boca Raton: CRC Press – Taylor & Francis Group. 6000 Broken Sound Parkway NW, Suite 300.

Turner, R. K., Burgess, D., Hadley, D., Coombes, E., & Jackson, N. (2007). A cost-benefit appraisal of coastal management realignment policy. *Global Environmental Change, 17*, 397–407.

UKCP09. (2009). *Data from the UK climate projection*. http://ukclimateprojections.defra.gov.uk/content/view/2068/517/

# Chapter 12
# Jellyfish Blooms and Their Impacts on Welfare Benefits: Recreation in the UK and Fisheries in Italy

**Maria Giovanna Palmieri, Marije Schaafsma, Tiziana Luisetti, Alberto Barausse, Amii Harwood, Antara Sen, and R.K. Turner**

## 12.1 Introduction

Jellyfish blooms may be regarded as an ecosystem shock, where the danger lies in the sudden outbreaks of jellyfish biomass, which may invoke changes and reactions in the ecosystem that are non-linear or affect multiple species (Daskalov et al. 2007). The main problem for a detailed assessment of the impacts of jellyfish outbreaks on ecosystem services provision is the paucity of jellyfish population datasets covering large temporal and spatial scales and the limited understanding of the role of jellyfish in ecosystems, the interaction with fish and other species

M.G. Palmieri (✉) • A. Harwood
Centre for Social and Economic Research on the Global Environment (CSERGE), School of Environmental Sciences, University of East Anglia, Norwich Research Park, Norwich NR4 7TJ, UK
e-mail: m.g.palmieri@uea.ac.uk

M. Schaafsma
Geography and Environment and Centre for Biological Sciences, University of Southampton, University Road, Southampton SO17 1BJ, UK

T. Luisetti
Centre for Environment Fisheries & Aquaculture Science (CEFAS), Pakefield Road, Lowestoft, Suffolk NR33 0HT, UK

A. Barausse
Environmental Systems Analysis Lab (LASA), University of Padova, via Marzolo 9, 35131 Padova, Italy

A. Sen
Affiliate of the Centre for Social and Economic Research on the Global Environment (CSERGE), School of Environmental Sciences, University of East Anglia, Norwich Research Park, Norwich NR4 7TJ, UK

R.K. Turner
School of Environmental Sciences, University of East Anglia, Norwich, UK

 
R.K. Turner, M. Schaafsma (eds.), *Coastal Zones Ecosystem Services*, Studies in Ecological Economics 9, DOI 10.1007/978-3-319-17214-9_12

populations, and the influence of human activities on the occurrence of blooms (Boero et al. 2008; Condon et al. 2012; Purcell et al. 2007). There is insufficient evidence to say under which conditions jellyfish blooms may cause irreversible ecosystem shifts, let alone to what extent these are caused by human pressures, as jellyfish also have a natural bloom-and-bust cycle following climatic patterns. The range of conditions at which ecosystems stay in a stable state is also unknown (see Fig. 2.4); shift are usually unpredictable (Scheffer et al. 2001).

In the absence of scientific knowledge about the physical, chemical and biological components of the marine ecosystem, economic valuation of the impacts of jellyfish blooms can only be based on some simplifying rules and assumptions to assess changes in ecosystem services provision, in combination with the exploration of different scenarios – alternative stable states of the ecosystem with prior and post levels of ecosystem services provision (Crépin et al. 2012).

This chapter provides a review of the occurrence, causes, and impacts of jellyfish blooms throughout the world. It then presents two case studies of jellyfish impacts in Europe and provides estimates of the potential welfare losses stemming from impacts of blooms on recreation in the UK and fisheries in Italy.

## 12.2 Jellyfish Blooms: Occurrence, Causes, and Impacts

### *12.2.1 Is a "Jellification" of Global Seas Taking Place?*

In the last decades extensive outbreaks of both indigenous and alien jellyfish have been recorded in several regions worldwide raising concern about a possible "jellification" of global seas (Jackson et al. 2001; Lynam et al. 2006; Attrill et al. 2007; Richardson et al. 2009). In the East China and Yellow Seas jellyfish blooms have become an annual event. Outbreaks of the giant jellyfish *Nemopilema nomurai* have taken place nearly each year since 2000 and blooms of other species, such as *Cyanea nozakii* and *Aurelia aurita*, have also increased during the same period (Dong et al. 2010). From Chinese waters jellyfish are carried northward into the Sea of Japan, where blooms of *Aurelia aurita* and *Nemopilema nomurai* also appear to have increased in the last decades (Uye 2008, 2010; Kawahara et al. 2006). In the Bering Sea the biomass of jellyfish, in particular the cnidarian *Chrysaora melanaster* (see Box 12.1), increased more than tenfold during the 1990s but declined after 2000 (Brodeur et al. 1999, 2002, 2008). The Northeast Atlantic has also recently witnessed an increasing trend in jellyfish abundance with outbreaks of a number of cnidarian species (Lilley et al. 2009; Licandro et al. 2010; Bastian et al. 2011; Lynam et al. 2011). In the southern North Sea and in Baltic waters the alien ctenophore *Mnemiopsis leidyi* has formed blooms along the coasts of the Netherlands, Belgium, and Denmark starting from 2006 (Faasse and Bayha 2006; Riisgård et al. 2012; Van Ginderdeuren et al. 2012). In the Mediterranean Sea outbreaks of this species have occurred in 2009 along the Mediterranean coasts of Italy, Spain, and Israel (Boero et al. 2009; Fuentes et al. 2010), while each summer since the

**Box 12.1. Glossary**

**Jellyfish**: Free-swimming gelatinous animals belonging to the phyla Ctenophora and Cnidaria.

**Ctenophora**: Invertebrate phylum, sometimes called comb jellies or sea gooseberries, which propel themselves through the sequential beating of their rows of cilia. They have colloblasts, which are cells that discharge a glue to ensnare preys. Ctenophores are holoplanktonic, lacking a benthic life stage and thus remaining in the plankton their entire life.

**Cnidaria**: Invertebrate phylum that contains animals, which vary in size from a few millimetres to a few metres. This phylum includes the "true jellyfish", which all produce structures called nematocysts and generally have alternating polyp and medusa life stages.

**Polyp**: The benthic stage of cnidarians with a general body plan of a cylindrical body and a ring of tentacles surrounding an oral opening.

**Medusa**: The mobile, bell-shaped stage of cnidarians that actively swim through muscular contraction of their bells.

**Nematocysts**: Stinging cells that are concentrated in the tentacles and mouth appendages of cnidarians and are used to poison or stun preys.

**Statoliths**: Calcium carbonate structures in the margin of the swimming bell of medusae that are used to sense gravity and so help in maintaining orientation.

(Adapted from Richardson et al. 2009)

mid-1980s swarms of another alien, the cnidarian *Rhopilema nomadica*, have appeared along the Levantine coast (Galil 2008). In addition to alien jellyfish, record abundances of indigenous species, including *Pelagia noctiluca*, *Aurelia aurita*, *Chrysaora hysoscella*, *Cothyloriza tuberculata*, and *Rhizostoma pulmo*, have been documented in the last decades in Mediterranean waters (CIESM 2001; Kogovšek et al. 2010; Brotz and Pauly 2012). *Mnemiopsis leidyi* was accidentally introduced in the Black Sea in the early 1980s and from there spread to the adjacent seas of Azov, Marmara, the Aegean and the Caspian (Costello et al. 2012). By the late 1980s and for many years the pelagic ecosystem of the Black Sea became a dead-end gelatinous food-web (Shiganova et al. 2001). In the northern Benguela off Namibia reports of blooms of *Chrysaora hysoscella* and *Aequorea forskalea* have been increasing since the 1990s (Lynam et al. 2006). In the Gulf of Mexico *Aurelia aurita* and *Chrysaora quinquecirrha* increased both in distribution and abundance from the mid-1980s for over 10 years (Graham 2001), while the alien *Phyllorhiza punctata* formed its first bloom in 2000 (Graham et al. 2003).

Some scientists argue that not enough information is available yet to confirm increasing trends of jellyfish blooms. Condon et al. (2012) suggest that the perception that outbreaks are increasing globally is based on reports of increases in a few regions and on media reports, while recent blooms may be simply part of long-term cycles in jellyfish populations. Condon et al. (2013) used all available long-term time series of annual jellyfish abundances in the global seas to test this latter hypothesis. The authors conclude that, although there has been a weak but significant overall increase in jellyfish abundance since the 1970s, the perceived global increase in jellyfish over the past decades coincides with the most recent rising phase of a pattern of decadal oscillations in jellyfish populations (i.e. natural bloom-and-bust cycles following climatic patterns). However, the results of the study also show that some coastal zones are experiencing enhanced jellyfish blooms, including the Sea of Japan, North Atlantic shelf regions, Barents Sea, Limfjorden (Denmark), and parts of the Mediterranean Sea, although jellyfish populations in these regions also exhibit decadal oscillations.

### *12.2.2 Environmental Perturbations Favouring Jellyfish Blooms*

Jellyfish are provided with a suite of attributes that enable them to survive in disturbed marine ecosystems and to rebound rapidly as conditions improve (Richardson et al. 2009). These attributes include fast growth rates, the ability to shrink when starved, and the capacity to fragment and regenerate (Richardson et al. 2009). Jellyfish are efficient, gluttonous and non-selective predators (Arai 1997). They can mature and reproduce quickly, both sexually and a-sexually (Arai 1997). Jellyfish can withstand poor environments more easily than fish, and can survive in polluted and hypoxic conditions (Purcell et al. 2001; Grove and Breitburg 2005).

Direct evidence linking jellyfish blooms to anthropogenic perturbations is lacking in most cases, but correlative evidence suggests the existence of links (Purcell 2012). Potential causes of abnormal jellyfish mass occurrence include overfishing, global warming, eutrophication, chemical pollution, the increase of artificial hard substrates, and the transport of exotic species (reviewed in Mills 2001; Hay 2006; Purcell et al. 2007; Richardson et al. 2009; Purcell 2012). Often, these causes co-occur and are mutually reinforcing (Purcell 2012).

Due to overfishing and ocean pollution, the natural predators of jellyfish, such as tuna, sharks, and sea turtles, are disappearing (Pauly et al. 1998). At the same time, resources are increasingly available to jellyfish as the abundance of zooplanktivorous fish, which compete with jellyfish for food, decreases (Purcell and Arai 2001; Daskalov et al. 2007).

Jellyfish outbreaks have been associated with variations in water mass and high salinity, as well as warm temperature, which influence jellyfish life cycles and reproductive output (Purcell 2005). Climate change may further increase the probabilities of jellyfish blooms. It may increase the availability of flagellates (single celled organisms, eaten by small zooplankton, on which jellyfish feed), lengthen the reproduction and growth season, as well as extend the spatial distribution of jellyfish poleward due to water temperature increases and shift the population distributions into currently colder areas (Richardson et al. 2009; Purcell 2012). At the same time, decreased oceanic $CO_2$ levels, leading to sea acidification, could impact on organisms that build shells or skeletons of calcium and thus favour the proliferation of gelatinous organisms (Attrill et al. 2007). Many jellyfish also have calcium statoliths for orientation, but it is unknown how acidification may impact on statoliths secretion (Purcell et al. 2007).

Eutrophication is another possible cause of jellyfish blooms. The high level of nutrients in eutrophied waters favours phytoplankton blooms and generally leads to greater biomass at all trophic levels, which implies more food for jellyfish polyps (Purcell et al. 2007). Eutrophication also causes complex changes in the food web, which can ultimately favour jellyfish outbreaks (Purcell et al. 2007). Another effect of eutrophication is the lowering of the oxygen levels, which jellyfish and polyps can sustain more easily than fish (Purcell et al. 2001; Grove and Breitburg 2005). Jellyfish can also better deal with turbidity and lower water clarity caused by eutrophication, as they do not need eyesight to hunt (Eiane et al. 1999).

Coastal and sea-shore development has created good places for jellyfish polyps to settle on, such as piers, marinas, aquaculture structures, oil platforms, and wind energy constructions (Duarte et al. 2013). Finally, alien jellyfish have invaded areas where they were transported to in ballast water or by hull fouling as polyps (Graham and Bayha 2007; Costello et al. 2012).

### *12.2.3 Impacts of Jellyfish Blooms on the Provision of Ecosystem Services*

The environmental change process in marine ecosystems may enhance jellyfish populations and blooms in the future, increasing the likelihood of negative jellyfish impacts on human activities. Currently, jellyfish blooms have negative impacts in a number of ways (reviewed in Purcell et al. 2007) but only a few economic estimates of these impacts are available.

Impacts on fisheries are the most frequently reported. These impacts arise because of the biological impacts of jellyfish on food webs and because of interference with fishing operations. Biological impacts derive from resource competition with fish and predation on fish eggs and juveniles (reviewed in Purcell and Arai 2001). This has been the case with the alien ctenophore *Mnemiopsis leidyi*, which contributed to the collapse of the anchovy fisheries in the Black Sea because of predation of *Mnemiopsis leidyi* on anchovy eggs and competition with anchovy for zooplankton (Shiganova et al. 2001). The collapse of the fishery caused significant economic losses estimated at hundreds of millions of US dollars over several decades (Knowler 2005). A decrease in the biomass of commercial fish species in association with an increase in jellyfish populations has also been observed elsewhere in the world (Lynam et al. 2005, 2006; Dong et al. 2010) but estimates of economic losses are not available. Jellyfish have been reported to interfere with fishing operations in a number of ways, including reduction in fish catches, clogging and bursting nets, requiring more labour to remove jellyfish from nets, increasing fish mortality due to nematocyst venom, causing painful stings to fishermen, displacing hauls to areas more distant from landing ports, and preventing fishermen from operating (Purcell et al. 2007 and references therein, Schiariti et al. 2008; Uye 2008; Nagata et al. 2009; Dong et al. 2010; Quiñones et al. 2013; Palmieri et al. 2014). In 2000 outbreaks of the alien *Phyllorhiza punctata* may have caused losses of up to USD 12 million[1] to the shrimp fishery of the northern Gulf of Mexico because of fouled fishing gear and harvest (Graham et al. 2003). In 2003 blooms of *Nemopilema nomurai* caused a loss in fishing revenue of approximately USD 18 million in just one of the 17 Japanese prefectures, where interferences of jellyfish with fishing operations were reported (Kawahara et al. 2006). Quiñones et al. (2013) estimated that in the austral summer 2008–2009 by-catch of *Chrysoara plocamia* caused losses of more than USD 200,000 to the Peruvian purse seiners of Ilo in only 35 days of fishing. Palmieri et al. (2014) estimated that economic losses due to reduced fish catches could amount to more than EUR eight million per year for the Italian Northern Adriatic trawling fleet if no additional fishing effort was made to mitigate losses.

Jellyfish blooms can also affect aquaculture. Jellyfish may damage shellfish and decapods (lobster, crab) culture and kill fish in cages through haemorrhage and suffocation (Purcell et al. 2007). A number of fish mass killings have taken place in the last years in salmon farms in Northern European countries. In 2002 almost one million salmon were lost at two farms in the Scottish Western Isles for a combined loss of around EUR three million.[2] In 2003 a jellyfish outbreak in Norway accounted for the death of over 600 t of farmed salmon (Heckmann 2004). In 2007 a jellyfish bloom in Northern Ireland caused a financial loss of a salmon farm stock of over EUR one million and a more recent bloom in 2013 caused other substantial losses.[3]

Jellyfish blooms have been reported to interfere with coastal power plant operations by clogging power plant intakes. In Israel the costs to remove the jellyfish from two power plants were estimated at almost USD 60,000 in just one summer (Galil 2008). In July 2008 over 4,000 t of *Aurelia aurita* were removed from the clogged intake screens of one power plant in China (Dong et al. 2010). In 2011 two nuclear reactors had to shut down for a few days at a Scottish power plant after an influx of jellyfish.[4]

Some jellyfish species interfere with recreational activities and have impacts on human health. Jellyfish stinging is a serious health problem along the coasts of some Asian and Indo-Pacific countries, where extremely venomous jellyfish are common and can cause death (Fenner and Williamson 1996; Burnett 2001). In other regions of the world, like the Mediterranean, stings from jellyfish are not lethal but may cause severe discomfort and pain and sometimes require medical treatment (Mariottini and Pane 2010; De Donno et al. 2009). In regions with popular touristic seaside resorts, stinging has sometimes occurred at epidemic levels. During the

[1] The original value, as well as other values in this section, have been corrected for inflation and converted to 2011 prices.

[2] http://news.bbc.co.uk/1/hi/scotland/2178959.stm

[3] http://www.irishtimes.com/news/ireland/irish-news/jellyfish-bloom-kills-thousands-of-farmed-salmon-off-co-mayo-1.1567468

[4] http://www.theguardian.com/environment/2011/jun/30/jellyfish-shut-nuclear-reactors-torness?guni=Article:in%20body%20link

summers 2006–2007, for instance, tens of thousands of bathers were stung by jellyfish on Spanish and French beaches (Galil 2008). It is unknown to what extent beach recreation has declined as a result of jellyfish presence but many coastal towns have taken remediating actions to protect the tourism industry. Spain has set out nets to mark out bathing areas and reduce the number of jellyfish to a minimum. According to local newspapers, the community near Mar Menor, which experiences annual blooms of two jellyfish species, spends around EUR 600,000 per year[5,6] to this effect. Similar systems have been deployed in Monaco and along the coasts of Cannes and Marseille (Galil 2008). It must be pointed out that the figures presented above on the costs of the investments to protect bathers from jellyfish are financial values and thus do not provide a complete picture of the economic welfare impact of jellyfish blooms on recreational activities (see Chap. 4). In addition to impairing swimming/bathing, jellyfish outbreaks can also impact on other recreational activities, like walking, when they lead to mass strandings and bad smell from decomposition on beaches.[7]

On the positive side, jellyfish provide some benefits to humans (reviewed in Purcell et al. 2007). Some species of jellyfish potentially enhance fisheries recruitment by providing shelter under their bells to fish juveniles, which feed on the prey and parasites of their hosts. Jellyfish are on the diet of many vertebrates, including commercially valuable fish species. Jellyfish are also used for human consumption. A number of jellyfish species have been historically fished in several Southeast Asian countries (Omori and Nakano 2001; Nishikawa et al. 2008; Kitamura and Omori 2010) and jellyfish fisheries have begun to be developed more recently in countries such as Australia, India, Turkey, Mexico, and the United States (Hsieh et al. 2001).

Jellyfish-derived products are being tested in the production of cosmetics and drugs. Jellyfish are believed to contain collagen, which moisturise the skin, and to cure rheumatoid arthritis and bronchitis (Hay 2006; Sugahara et al. 2006). They provide Aequorin and Obelin, which are green fluorescent proteins that are for instance used as biomarkers in biomedical research.

Among the potential benefits of jellyfish is carbon uptake from the atmosphere. According to Condon et al. (2011), who looked at *Mnemiopsis leidyi*, jellyfish may be net up-takers of oceanic carbon. Evidence, however, is scarce and not conclusive (Brotz et al. 2011).

Finally, jellyfish have a recreational and educational value. There is a small niche for jellyfish recreation, where diving for luminescent jellyfish is possible or where the species do not sting, as for example in the Jellyfish Lake in the Palau Archipelago (Dawson et al. 2001). Jellyfish are also a popular attraction in many marine aquaria worldwide.

[5] http://murciatoday.com/300-tons-of-jellyfish-extracted-from-the-mar-menor-in-the-last-8-days_12646-a.html

[6] http://murciatoday.com/jellyfish-nets-ready-and-waiting-in-the-mar-menor_17156-a.html

[7] http://iltirreno.gelocal.it/pisa/cronaca/2012/04/24/news/colpa-delle-meduse-i-cattivi-odori-apparsi-sul-litorale-1.4416394

# 12.3 Case Study in the UK: Recreation

## 12.3.1 Introduction

In the first case study, we discuss the potential impacts on tourism of jellyfish blooms along the English coastline. The abundance of jellyfish appears to have increased in the last two decades in the seas surrounding the UK due to climate variation and possibly to changes in food web structure (Attrill et al. 2007; Gibbons and Richardson 2009; Lynam et al. 2011). Climate change is expected to contribute to a further increase in jellyfish frequency over the next century (Attrill et al. 2007) and the occurrence of warm water species, such as *Pelagia noctiluca*, might become more frequent (Licandro et al. 2010). Moreover, human activities, such as maritime transport, may favour the introduction of alien species.

Along the English coast, beach recreation is an important activity for local, national and international visitors (Sen et al. 2011). Jellyfish outbreaks could therefore have a wide impact. Little information is available regarding the number of tourists that visit beaches along the English coast, and there are no studies that estimate the direct effect of jellyfish blooms on recreation along the English coastline.

## 12.3.2 Data and Methodology

To approximate the economic damage cost that jellyfish blooms may have in the future, we use the analysis produced for the UK National Ecosystem Assessment (NEA) on recreation (UK-NEA 2011). The spatially explicit approach of the UK-NEA is described in detail in Sen et al. (2014), who modelled the non-market value of open-air recreation throughout England. The model is based on a large survey about recreational behaviour among households in England (Natural England 2010). The model predicting annual visitor numbers takes into account a wide range of spatial characteristics, including habitats, population and accessibility. One of the findings of this model is that the number of trips to coastal areas is higher than for most other types of land cover, including grasslands, mountains, or woodlands. The model is combined with a meta-analysis on the value per recreational trip across different types of habitats. By multiplying the estimated number of visits by the value per trip, an estimate of the total annual value of visits to sites is obtained. The analysis is performed using GIS at a 1 $km^2$ scale.

We focus our analysis on the English part of the ICES fishing areas: zones VIIf, VIIe, VIId and IVc.[8] From the predicted visitor numbers to all 1 $km^2$ cells in the UK,

[8] ICES areas VIIe and VIIf include Cornwall and Isles of Scilly, Devon, Dorset, Gloucestershire, North Somerset, Plymouth, Poole, Somerset, South Gloucester, and Torbay. VIId includes Bournemouth, Brighton and Hove, Bristol, East Sussex, Hampshire, Portsmouth, Southampton

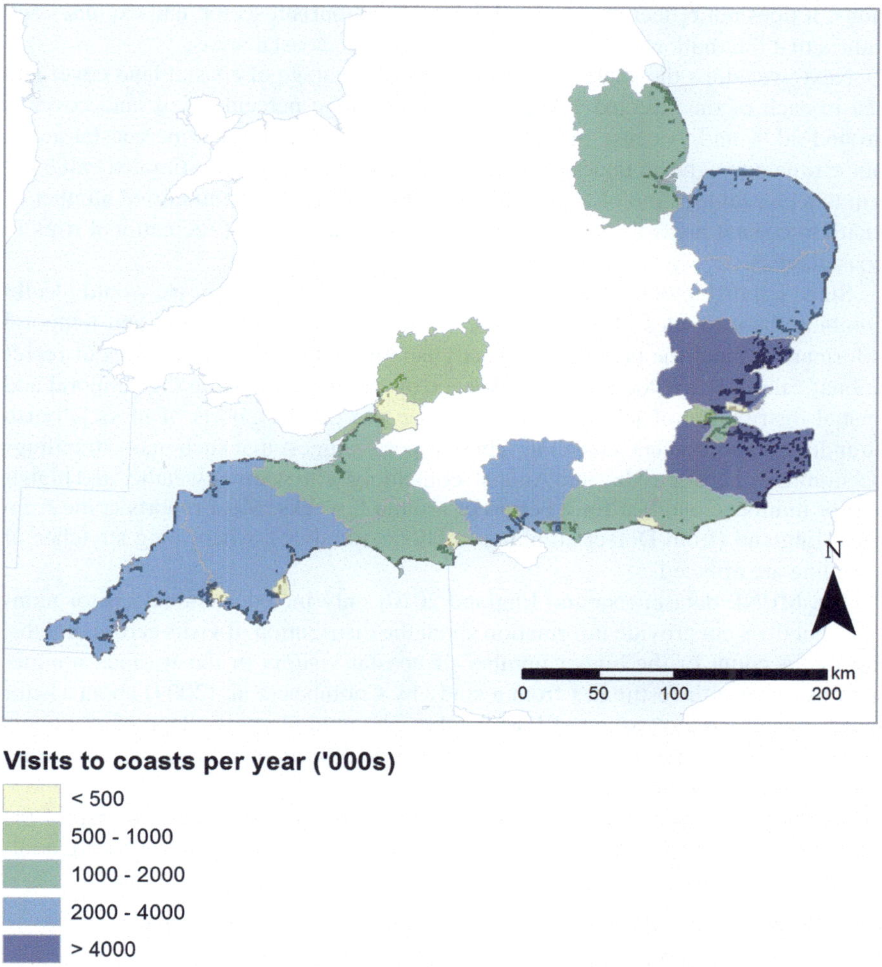


**Fig. 12.1** Annual number of visits to coastal areas

we select the cells in the 26 counties in our study area (England), which contain some coastal land cover or sea according to the NEA definition of habitat types, within a 10 km distance from the coastline.

Figure 12.1 presents a map of the estimated annual number of visits to coastal locations. The map shows that Cornwall, Devon, Hampshire, Norfolk, Essex and Kent generate high visitor numbers. They will therefore generate higher benefits, representing the value that English households attach to recreation at coastal loca-

and West Sussex. IVc includes Essex, Kent, Lincolnshire, Medway, Norfolk, Southend-on-Sea, Suffolk and Thurrock.

tions. It does not reflect the financial value of the tourism sector, and excludes any values that international visitors may attach to these coastal areas.

Next, we adjust the visitor numbers by the percentage of coastal land cover and sea in each of the selected 1 $km^2$ cells. The average percentage of land cover is around 20 % and, because the NEA land cover map excludes some coastal areas, our estimate may be an underestimate. Finally, we multiply the estimated value per trip to a coastal location of EUR 4.67 (Sen et al. 2011)[9] by the modified number of visits to coastal areas to get an estimate of the total value of recreational trips to coastal areas.

Since jellyfish blooms have a spatial and temporal dimension, we would ideally combine the estimates of annual value of coastal visits with spatial and temporal information about the probability of jellyfish blooms to assess the losses of recreational value due to blooms. In the absence of monitoring data on the temporal and spatial distribution of jellyfish outbreaks, we use public reports of mass jellyfish strandings along the UK coastline. These reports suggest that such mass strandings occur mostly between May and August, coinciding with warmer weather and higher visitor numbers, and last for a period of around 2 weeks. Most reports come from West-England (from Dorset all along to Gloucestershire), where long stretches of coastline are affected.

The MENE dataset (Natural England 2010) only includes annual visitor numbers, but does not provide information about the distribution of visits across months. To take account of the higher number of coastal visitors in the warmer summer months, we use the estimates from a study by Coombes et al. (2009) about visitor numbers along the coastline of East Anglia. The monthly percentage of visitors is higher for the summer months. We assume that visitors in other areas have a similar distribution across the year.

To assess the economic loss of jellyfish blooms in the absence of a map of the probability of the spread and timing of such events, we make the following additional assumptions:

- Jellyfish blooms affect coastal visits through mass beach strandings;
- Mass jellyfish strandings create such a stink that the trip creates no net benefit to the visitor, who may also consider health risks, i.e. the value per trip is set equal to zero;
- A typical mass jellyfish stranding lasts 2 weeks;
- Mass jellyfish strandings affect large areas. Therefore, we combine 26 different counties into three areas roughly according to the ICES fishing areas IVc, VIId, and VIIe and VIIf (England only);
- All sea and coastal land cover cells are equally and simultaneously affected, i.e. within the ICES areas we assume that all beaches are equally affected.

[9] Note that this is a fixed value per trip and does not account for variation in values due to seasonality.

**Table 12.1** Recreational benefits lost to 2-week mass jellyfish strandings along the English coastline (EUR*1000, 2011 prices)

| ICES area | May | June | July | August |
|---|---|---|---|---|
| VIIe+VIIf | 527 | 717 | 965 | 1,060 |
| VIId | 414 | 563 | 757 | 832 |
| IVc | 761 | 1,036 | 1,392 | 1,529 |
| *Total* | 1,703 | 2,317 | 2,903 | 3,421 |

### *12.3.3 Results*

The recreational benefits lost across different ICES areas for different summer months based on these assumptions are summarised in Table 12.1. The table shows that, under these assumptions, a jellyfish swarm affecting all English coastal waters in August would imply a loss of recreational values of over EUR 3.4 million (2011 prices). However, such large scale swarms have not been registered in the past, which needs to be taken into account when interpreting the figures in Table 12.1.

For a widespread 2-week jellyfish outbreak, the loss of recreational values in the IVc area, which borders the North Sea, would be highest, varying between EUR 0.8 and 1.5 million. The VIIe and VIIf areas, bordering the Bristol Channel, Celtic Sea, and eastern part of the English Channel, would incur lower value losses, and the smallest zone VIId reflecting the western part of the English Channel would see lowest value losses, according to these estimates. Note, however, that we cannot account for the probability of jellyfish blooms and these estimates do not reflect expected values, but historical records suggest that jellyfish are more likely to be found in ICES areas VIIe and VIIf than in the other two ICES areas.

In summary, this section presents a methodological approach that can be used when stated preference surveys are not applicable to get a first order-of-magnitude estimate of potential losses in social welfare related to the impacts of jellyfish abundance on recreational benefits along the English shoreline. It could be improved when more spatial information is available about the scale and spatial and temporal distribution of jellyfish blooms and visitation rates, as well as information to support and refine the assumptions of visitor reactions to these blooms.

## 12.4 Case study in Italy: Fisheries

### *12.4.1 Introduction*

The Northern Adriatic (NA) Sea is one of the most exploited Mediterranean fishing grounds (Barausse et al. 2009), although the high primary productivity of the ecosystem has clearly decreased since the late 1990s (Mozetic et al. 2010). Starting from the end of the 1980s, plankton, fish and invertebrate communities in the NA underwent abrupt changes, which were collectively identified as a regime shift in the ecosystem, probably driven by the interaction of different pressures, such as climate change, which caused variations in water temperature and circulation; reduced nutrient inputs from river catchments; anoxic phenomena; overexploitation of fishery resources; the crash of the stock of anchovy, a species which plays a key role in the food web; and the 10-year long bloom of the jellyfish *Pelagia noctiluca* (Barausse et al. 2011). This species competed with small pelagics for zooplankton and predated upon fish eggs, larvae and even adults, possibly stimulating the aforementioned anchovy population collapse and altering ecosystem functioning (Boero and Bonsdorff 2007; Conversi et al. 2010; Kogovšek et al. 2010; Barausse et al. 2011). Apart from the *Pelagia*'s massive bloom, from the early 1980s the NA has experienced blooms of a number of other jellyfish species, whose occurrence appears to have increased in recent decades (Kogovšek et al. 2010). In this case study we discuss the links between welfare benefits (i.e. fishery landings), jellyfish blooms, and anthropogenic pressures in the NA ecosystem.

### *12.4.2 Data and Methodology*

We use the Ecopath with Ecosim (EwE) modelling suite (see Chap. 3) to investigate the links between fishery landings, jellyfish blooms, and anthropogenic pressures, such as nutrient enrichment, in the NA ecosystem.

Scenarios of jellyfish outbreaks are modeled using an Ecosim simulation (Christensen and Walters 2004) based on the Ecopath trophic network model of the NA Sea described in Barausse et al. (2009). Ecosim can simulate the variation in the biomass of food web compartments over time based on processes such as changes in system productivity, fishing mortality, predator-prey interactions, migration fluxes, and biological invasions. Here, Ecosim (version 5.1) is used to assess how the NA ecosystem and particularly fisheries respond to sudden jellyfish blooms or invasions triggered by non-trophic causes (e.g. some unknown factor such as climate), which are simulated by forcing jellyfish abundance in the model. The parameter values of the Ecosim model, such as vulnerabilities, were previously calibrated by fitting the model to time series over the period 1996–2006 (Alberto Barausse, University of Padova, unpublished data). Four scenarios (S1 − 4) are run, all depicting the effects of an abrupt increase in jellyfish biomass (which is forced in the model to simulate a sudden bloom or invasion, as explained above):

- S1: 3-year jellyfish bloom, constant primary production
- S2: 3-year jellyfish bloom, 10 % decrease in primary production from 2006 to 2020
- S3: 10-year jellyfish bloom, constant primary production
- S4: 10-year jellyfish bloom, 10 % decrease in primary production from 2006 to 2020

Blooms are started at the end of the fitting period, i.e. 2007, and are assumed to last 3 years (scenarios S1 and S2), based on what has recently happened in the NA Sea for *Pelagia noctiluca* during the 2000s, or, alternatively, 10 years (scenarios S3 and S4), based on the intense outbreak of the same species which took place in the ecosystem in the period 1977–1986 (Kogovsek et al. 2010). *Pelagia noctiluca* is one of the most ecologically important jellyfish species in the ecosystem; it eats zooplankton, fish eggs, larvae and juveniles and can even kill adult fishes (Fernando Boero, Università del Salento, pers. comm.). Based on data in Malej and Malej (2004), Barausse et al. (2009) and Kogovsek et al. (2010), jellyfish biomass during the bloom is assumed to be twelve times as high as the biomass in the baseline Ecopath model; in the fourth (S1, S2) or the eleventh (S3, S4) year after the start of the bloom, jellyfish biomass is forced back to the baseline Ecopath value to simulate the end of the outbreak. Such bloom magnitude is likely to be conservative and to underestimate the real impact of gelatinous plankton outbreaks because reported estimates of jellyfish biomass in the NA Sea mostly refer to bloom conditions and the actual increase in jellyfish biomass during blooms should be, therefore, much higher.

To evaluate the ecological impact of the jellyfish bloom on fish landings, the model is run until 2020, the year when a Good Environmental Status should be achieved in Europe's seas according to the Marine Strategy Framework Directive (2008/56/EC). In the modelling scenarios S1 and S3, fishing effort, fishing mortalities and primary productivity in the ecosystem are kept equal to the 2006 values over 2007–2020, while in the scenarios S2 and S4 a 10 % linear decrease in phytoplanktonic primary productivity from 2006 to 2020 is simulated to mirror the current oligotrophication of the system (Mozetic et al. 2010), which is expected to affect the NA fisheries (Barausse et al. 2011). The impact of jellyfish on landings is evaluated in each scenario by comparing landings in 2020 with the landings predicted in that same year by a "reference" scenario which is identical from a modelling point of view (e.g. same changes in primary production and fishing pressure), except that no jellyfish outbreak is simulated.

To assess the welfare impact of the bloom on the NA Italian fisheries, we estimate the change in revenue based on the variation in landings. As the data in the model refer to five Italian fisheries sorted according to the fishing gear and one pooled Slovenian-Croatian fishery (as described in Barausse et al. 2009), we extract the data pertaining to the Italian landings. After collecting data on the 2011 prices of landings by fishing fleet in the three Italian regions of the NA (Veneto, Emilia Romagna, and Friuli Venezia Giulia), we calculate the mean prices of landings weighted on the basis of the quantities landed (IREPA 2012). The impact of the jellyfish bloom in terms of lost revenue is estimated by multiplying the price per kg per fishing fleet by the variation in landings in relation to the four scenarios.

### *12.4.3 Results*

Table 12.2 reports the percentage change in landings in 2020 in the different scenarios, with respect to reference scenarios where no jellyfish bloom takes place. Interestingly, the results suggest that jellyfish blooms always have overall negative impacts on fisheries, since in all scenarios the blooms cause a decrease in total landings in 2020 with respect to the reference scenarios, a decrease of about 0.5 % in the case of the 3-year blooms and of about 2.3 % in the case of the 10-year blooms. These figures show that the response of fisheries to jellyfish blooms is disproportionately more negative in the case of the longer-lasting blooms, as such blooms (which are 3.3 times longer than the shorter-lasting ones) cause a decrease in landings which is 4.6 times stronger than the one caused by the shorter-lasting blooms. The model responses are not particularly sensitive to simulated changes in primary

**Table 12.2** Changes in fishery landings (%) in 2020 due to the jellyfish bloom, for each model scenario. Changes were calculated with respect to the landings simulated in 2020 with the same trends in all other forcing functions (fishing mortality, primary productivity) as the given scenario except that no jellyfish bloom was simulated. Variations in landings are reported according to the fished group and to total fishery landings in the basin

| | Scenario | | | |
|---|---|---|---|---|
| | S1 | S2 | S3 | S4 |
| Food web group | (%) | (%) | (%) | (%) |
| Sharks | 0.3 | 0.3 | 1.9 | 1.2 |
| Rays | 0.0 | 0.0 | 1.8 | 2.0 |
| European hake | 0.0 | 0.0 | 2.0 | −0.2 |
| Zoobenthivorous fish – hard bottom | 0.0 | 0.0 | −1.7 | −1.3 |
| Zoobenthivorous fish – soft bottom | 0.0 | 0.0 | −0.7 | −0.6 |
| Mackerel | 1.2 | 1.3 | 9.6 | 6.6 |
| Horse mackerel | 1.0 | 1.0 | 7.4 | 5.3 |
| Other small pelagics | 0.0 | 0.0 | −0.6 | 0.5 |
| Anchovies | 0.5 | 0.5 | 1.6 | 1.2 |
| Sardines | −6.4 | −6.5 | −29.5 | −29.7 |
| Nectobenthic zooplanktivorous fish | −0.1 | −0.2 | 1.8 | 2.5 |
| Omnivorous fish | −1.6 | −1.6 | −7.6 | −5.7 |
| Benthic piscivorous fish | −0.3 | −0.3 | 4.7 | 2.5 |
| Flatfishes | −0.5 | −0.6 | −2.5 | −1.9 |
| Squids | −2.8 | −3.0 | −7.0 | −10.0 |
| Benthic cephalopods | −0.2 | −0.2 | −1.2 | −0.5 |
| Macro-crustaceans | 0.0 | 0.0 | −0.1 | 0.0 |
| Mantis shrimp | 0.0 | 0.0 | −0.3 | −0.4 |
| Commercial bivalves | 0.0 | 0.0 | −0.1 | 0.0 |
| Gastropods | 0.0 | 0.0 | −0.2 | −0.1 |
| Filter feeding invertebrates | 0.0 | 0.0 | 0.7 | 0.8 |
| *Total landings* | *−0.5* | *−0.5* | *−2.3* | *−2.4* |

production. Only few differences can be appreciated between total landings in scenarios S1 and S2, and in scenarios S3 and S4 (but some exceptions can be observed for single groups in S3 and S4), suggesting that in general a reduction in system primary productivity does not act synergistically with jellyfish outbreaks in reducing landings.

For all modelled food web groups, the response to jellyfish blooms is (often much) weaker in the case of the scenarios simulating the shorter-lasting jellyfish bloom. However, even a 3-year bloom causes a decrease of about 6.5 % in the landings of sardine, which is a key commercial species in the NA and also plays an important trophic role in the ecosystem (Barausse et al. 2009). In the case of the 10-year bloom, sardine fisheries are impacted heavily with decreases in landings of about 30 %. Instead, anchovy, another commercially and ecologically important species, gains some benefits from the jellyfish outbreaks, probably due to reduced competition for zooplankton with sardine, and its landings show a slight increase in all scenarios. In general, responses to jellyfish blooms vary across groups in a complex manner, with landings of medium-low trophic level groups feeding on or a few trophic connections away from zooplankton (by far the main food of jellyfish) being most strongly affected. For example, landings of mackerel and horse mackerel increase, since these two groups mostly feed on zooplankton and small pelagic fish such as anchovy, while squid catches decrease possibly due to food competition for small pelagics. Interestingly, landings in benthic piscivorous fish decrease slightly in the presence of a short jellyfish bloom, but increase markedly when a 10-year bloom is simulated, and moreover the decrease rate depends clearly on the simulated trend in primary production, suggesting that complex food web interactions define their response.

Looking at the response of different fleets to the jellyfish bloom, landings from all fleets decrease (data not shown). The Italian fleets account for around 60–70 % of the total reduction in catches landed in the NA region across the four scenarios.

Table 12.3 reports the changes in Italian landings and revenues (undiscounted and in 2011 prices) due to the jellyfish bloom for each model scenario. The strongest response is observed for the mid-water trawling fleet. The reduction in catches by this fleet accounts for around 90 % of the total reduction in landings across scenarios. However, the revenue losses account for only around 50–60 % of the total. This is because the mid-water trawling fleet catches large amounts of a limited number of low value species (Gramitto et al. 2010), some of which are heavily impacted by the jellyfish bloom, such as sardine and common mullets (omnivorous fish group). The other trawling fleets catch smaller quantities of a higher number of species, many of which of high commercial value (Gramitto et al. 2010), and the different magnitude of their losses may depend on the diversification of their catches. While the otter trawling fleet targets some high value species heavily impacted by the bloom, such as squids and soles (flatfishes group), the beam trawling fleet, in addition to impacted species, also targets some that are marginally impacted, such as gastropods and bivalves. The latter is the main target of hydraulic dredges, which appear to be impacted negligibly. Heavy losses are registered for the

**Table 12.3** Changes in Italian NA fishery landings (t) and revenues (€, 2011 prices) in 2020 due to the jellyfish bloom, for each model scenario. Changes were calculated with respect to the landings simulated in 2020 with the same trends in all other forcing functions (fishing mortality, primary productivity) as the given scenario except that no jellyfish bloom was simulated. Changes are reported by fishing fleet and according to total landings by all Italian NA fleets. Revenue values are undiscounted. Values reported in parentheses are percentage changes

| | Scenario | | | | | | | |
|---|---|---|---|---|---|---|---|---|
| | S1 | | S2 | | S3 | | S4 | |
| Fishing fleet | (t) | (€) | (t) | (€) | (t) | (€) | (t) | (€) |
| Hydraulic dredges | −0.1 | −410 | −0.2 | −614 | −4.8 | −15,462 | 1.5 | 4,915 |
| | | (0.0) | | (0.0) | | (−0.1) | | (0.0) |
| Beam trawling | −4.9 | −31,046 | −5.0 | −31,248 | −30.8 | −193,738 | −18.4 | −115,718 |
| | | (−0.1) | | (−0.1) | | (−0.5) | | (−0.3) |
| Otter trawling | −14.3 | −89,914 | −14.0 | −88,301 | −49.6 | −312,682 | −43.6 | −274,378 |
| | | (−0.2) | | (−0.2) | | (−0.6) | | (−0.6) |
| Mid water trawling | −220.5 | −198,432 | −199.4 | −179,482 | −1,232.6 | −1,109,376 | −1,294.0 | −1,164,643 |
| | | (−0.4) | | (−0.4) | | (−2.1) | | (−2.5) |
| Other and artisanal fisheries | −11.6 | −86,880 | −11.9 | −89,520 | −57.5 | −431,280 | −41.5 | −311,520 |
| | | (−0.5) | | (−0.5) | | (−2.3) | | (−1.7) |
| *Total landings* | *−251.4* | *−406,682* | *−230.5* | *−389,165* | *−1,375.3* | *−2,062,538* | *−1,396.0* | *−1,861,344* |
| | *(−0.3)* | *(−0.2)* | *(−0.3)* | *(−0.2)* | *(−1.7)* | *(−1.1)* | *(−1.9)* | *(−1.1)* |

small scale fisheries, which target heavily impacted species of high value, such as soles (Gramitto et al. 2010).

If we look at the overall reduction in the Italian NA landings, we can see that a 10-year bloom could cause a decrease in landings of around 2 %. This may have repercussions at the national level in terms of decreased seafood supply, especially in the case of the heavily impacted sardine, as around 40 % of the national production of this species originates from the Adriatic basin (Mulazzani et al. 2012), but it is also the case for other species, such as soles, which are more abundant in the NA region compared to other Mediterranean fishing grounds (Grati et al. 2013).

In terms of revenues, a 10-year bloom could entail revenue reductions to the Italian NA fisheries of around EUR two million (undiscounted and in 2011 prices). Decreasing revenues, such as those we describe here, could put pressure on the financial viability of fishing enterprises and affect the ability of the fishing industry to support, through employment and incomes, the economy and community cohesion of the small coastal communities of the region.

## 12.5 Conclusions

The case studies described in this chapter show that jellyfish blooms can have marked effects on recreation and fisheries, entailing considerable losses of welfare benefits. This result confirms the evidence collected in the literature on the negative impacts of jellyfish outbreaks, which include not only impacts on recreation and fisheries but also on other benefits, such as other production activities (e.g. aquaculture, energy production) and human health.

These results warrant a consideration of increased efforts towards the monitoring and control of jellyfish blooms. To this end, it is necessary to improve the scientific knowledge about trends in jellyfish populations through long-term monitoring programmes, which could provide indicators of jellyfish outbreaks (Condon et al. 2012), as in the case of the invasion of alien species (Stohlgren and Schnase 2006). It will be necessary to improve the understanding of the role that jellyfish play in ecosystems, as there is a lack of knowledge of the biology and ecology of these organisms (Boero et al. 2008). Furthermore, it would be good to investigate the influence of human activities on the occurrence of blooms through the development of models including system stressors (e.g. fishing, eutrophication, global warming) to assess their relative importance and explore ecosystem resilience (Richardson et al. 2009).

Until a higher level of understanding of jellyfish blooms is gained, we need to deal with their impacts based on current information. As long as there is uncertainty about what constitutes ecological threshold points, careful management that keeps ecological changes within some safe minimum standards should be advocated unless the social costs are unacceptable (Crowards 1998; Perrings 2001). Harmful algal blooms (HABs) may provide a model for the management of jellyfish blooms. Past controversies on HABs trends have been overcome through pragmatic discussions of the management of their impacts and more resources are now dedicated to the monitoring of HABs and to their control (Brotz and Pauly 2012).

# References

Arai, M. N. (1997). *A functional biology of scyphozoa*. London: Chapman & Hall.

Attrill, M. J., Wright, J., & Edwards, M. (2007). Climate-related increases in jellyfish frequency suggest a more gelatinous future for the North Sea. *Limnology and Oceanography, 52*, 480–485.

Barausse, A., Duci, A., Mazzoldi, C., Artioli, Y., & Palmeri, L. (2009). Trophic network model of the Northern Adriatic Sea: Analysis of an exploited and eutrophic ecosystem. *Estuarine, Coastal and Shelf Science, 83*(4), 577–590.

Barausse, A., Michieli, A., Riginella, E., Palmeri, L., & Mazzoldi, C. (2011). Long-term changes in community composition and life-history traits in a highly exploited basin (northern Adriatic Sea): The role of environment and anthropogenic pressures. *Journal of Fish Biology, 79*, 1453–1486.

Bastian, T., Stokes, D., Kelleher, J. E., Hays, G. C., Davenport, J., & Doyle, T. (2011). Fisheries bycatch data provide insights into the distribution of the mauve stinger (*Pelagia noctiluca*) around Ireland. *ICES Journal of Marine Science, 68*(3), 436–443.

Boero, F., & Bonsdorff, E. (2007). A conceptual framework for marine biodiversity and ecosystem functioning. *Marine Ecology, 28*, 134–145.

Boero, F., Bouillon, J., Gravili, C., Miglietta, M. P., Parsons, T., & Piraino, S. (2008). Gelatinous plankton: Irregularities rule the world (sometimes). *Marine Ecology Progress Series, 356*, 299–310.

Boero, F., Putti, M., Trainito, E., Prontera, E., Piraino, S., & Shiganova, T. A. (2009). First records of *Mnemiopsis leidyi* (Ctenophora) from the Ligurian, Thyrrhenian and Ionian Seas (Western Mediterranean) and first record of *Phyllorhiza punctata* (Cnidaria) from the Western Mediterranean. *Aquatic Invasions, 4*(4), 675–680.

Brodeur, R. D., Mills, C. E., Overland, J. E., Walters, G. E., & Schumacher, J. D. (1999). Evidence for a substantial increase in gelatinous zooplankton in the Bering Sea, with possible links to climate change. *Fisheries Oceanography, 8*, 296–306.

Brodeur, R. D., Sugisaki, H., & Hunt, G. L., Jr. (2002). Increases in jellyfish biomass in the Bering Sea: Implications for the ecosystem. *Marine Ecology Progress Series, 233*, 89–103.

Brodeur, R. D., Decker, M. B., Cianelli, L., Purcell, J. E., Bond, N. A., Stabeno, P. J., Acuna, E., & Hunt, G. L., Jr. (2008). Rise and fall of jellyfish in the eastern Bering Sea in relation to climate regime shifts. *Progress in Oceanography, 77*, 103–111.

Brotz, L., & Pauly, D. (2012). Jellyfish populations in the Mediterranean Sea. *Acta Adriatica, 53*(2), 213–232.

Brotz, L., Lebrato, M., Robinson, K. L., Sexton, M., Sweetman, A. K., Pitt, K., & Condon, R. (2011). Implications of increased carbon supply for the global expansion of jellyfish blooms. *Limnology & Oceanography Bulletin, 20*, 38–39.

Burnett, J. W. (2001). Medical aspects of jellyfish envenomation: Pathogenesis, case reporting and therapy. *Hydrobiologia, 451*, 1–9.

Christensen, V., & Walters, C. J. (2004). Ecopath with Ecosim: Methods, capabilities and limitations. *Ecological Modelling, 172*, 109–139.

CIESM. (2001). *Gelatinous plankton outbreaks: Theory and practice*. CIESM Workshop Series 14. Monaco: CIESM, 112pp.

Condon, R. H., Steinberg, D. K., Del Giorgio, P. A., Bouvier, T. C., Bronk, D. A., Graham, W. M., & Ducklow, H. W. (2011). Jellyfish blooms result in a major microbial respiratory sink of carbon in marine systems. *Proceedings of the National Academy of Sciences of the United States of America, 108*(25), 10225–10230.

Condon, R. H., Graham, W. M., Duarte, C. M., Pitt, K. A., Lucas, C. H., Haddock, S. H., Sutherland, K. R., Robinson, K. L., Dawson, M. N., Decker, M. B., Mills, C. E., Purcell, J. E., Malej, A., Mianzan, H., Uye, S.-i., Gelcich, S., & Madin, L. P. (2012). Questioning the rise of gelatinous zooplankton in the world's oceans. *Bioscience, 62*(2), 160–169.

Condon, R. H., Duarte, C. M., Pitt, K. A., Robinson, K. L., Lucas, C. H., Sutherland, K. R., Mianzan, H., Bogeberg, M., Purcell, J. E., Decker, M. B., Uye, S. I., Madin, L. P., Brodeur, R. D., Haddock, S. H., Malej, A., Parry, G. D., Eriksen, E., Quinones, J., Acha, M., Harvey, M., Arthur, J. M., & Graham, W. M. (2013). Recurrent jellyfish blooms are a consequence of global oscillations. *Proceedings of the National Academy of Sciences of the United States of America, 110*(3), 1000–1005.

Conversi, A., Fonda Umani, S., Peluso, T., Molinero, J. C., Santojanni, A., & Edwards, M. (2010). The Mediterranean sea regime shift at the end of the 1980s, and intriguing parallelisms with other European basins. *PLoS One, 5*(5), e10633.

Coombes, E. G., Jones, A. P., & Sutherland, W. J. (2009). The implications of climate change on coastal visitor numbers: A regional analysis. *Journal of Coastal Research, 25*(4), 981–990.

Costello, J. H., Bayha, K. M., Mianzan, H. W., Shiganova, T. A., & Purcel, J. E. (2012). Transitions of *Mnemiopsis leidyi* (Ctenophora: Lobata) from a native to an exotic species: A review. *Hydrobiologia, 690*, 21–46.

Crépin, A.-S., Biggs, R., Polasky, S., Troell, M., & de Zeeuw, A. (2012). Regime shifts and management. *Ecological Economics, 84*, 15–22.

Crowards, T. M. (1998). Safe minimum standards: costs and opportunities. *Ecological Economics, 25*(3), 303–314.

Daskalov, G. M., Grishin, A. N., Rodionov, S., & Mihneva, V. (2007). Trophic cascades triggered by overfishing reveal possible mechanisms of ecosystem regime shifts. *Proceedings of the National Academy of Sciences of the United States of America, 104*(25), 10518–10523.

Dawson, M. N., Martin, L. E., & Penland, L. K. (2001). Jellyfish swarms, tourists, and the Christ-child. *Hydrobiologia, 451*, 131–144.

De Donno, A., Idolo, A., & Bagordo, F. (2009). Epidemiology of jellyfish stings reported to summer health centres in the Salento peninsula (Italy). *Contact Dermatitis, 60*(6), 330–335.

Dong, Z., Liu, D., & Keesing, J. K. (2010). Jellyfish blooms in China: Dominant species, causes and consequences. *Marine Pollution Bulletin, 60*(7), 954–963.

Duarte, C. M., Pitt, K. A., Lucas, C. H., Purcell, J. E., Uye, S. I., Robinson, K., Brotz, L., Decker, M. B., Sutherland, K. R., Malej, A., Madin, L., Mianzan, H., Gili, J. M., Fuentes, V., Atienza, D., Pages, F., Breitburg, D., Malek, J., Graham, W. M., & Condon, R. H. (2013). Is global ocean sprawl a cause of jellyfish blooms? *Frontiers in Ecology and the Environment, 11*(2), 91–97.

Eiane, K., Aksnes, D. L., Bagøien, E., & Kaartvedt, S. (1999). Fish or jellies – A question of visibility? *Limnology and Oceanography, 44*(5), 1352–1357.

Faasse, M. A., & Bayha, K. M. (2006). The ctenophore *Mnemiopsis leidyi* A. Agassiz 1865 in coastal waters of the Netherlands: An unrecognized invasion? *Aquatic Invasions, 1*(4), 270–277.

Fenner, P. J., & Williamson, J. A. (1996). Worldwide deaths and severe envenomation from jellyfish stings. *Medical Journal of Australia, 165*, 658–661.

Fuentes, V. L., Angel, D. L., Bayha, K. M., Atienza, D., Edelist, D., Bordehore, C., Gili, J. M., & Purcell, J. E. (2010). Blooms of the invasive ctenophore *Mnemiopsis leidyi* span the Mediterranean Sea in 2009. *Hydrobiologia, 645*, 23–37.

Galil, B.S. (2008). The price of change: The economic impacts of alien species and jellyfish outbreaks in the Mediterranean Sea. In: *Economic valuation of natural coastal and marine ecosystems*. CIESM Workshop Monographs, No. 37. Monaco: CIESM, 112 pp.

Gibbons, M. J., & Richardson, A. J. (2009). Patterns of pelagic cnidarian abundance in the North Atlantic. *Hydrobiologia, 616*, 51–65.

Graham, W. M. (2001). Numerical increases and distributional shifts of *Chrysaora quinquecirrha* (Desor) and *Aurelia aurita* (Linné) (Cnidaria: Scyphozoa) in the northern Gulf of Mexico. *Hydrobiologia, 451*, 97–111.

Graham, W. M., & Bayha, K. M. (2007). Biological invasions by marine jellyfish. *Ecological Studies, 193*(5), 239–255.

Graham, W. M., Martin, D. L., Felder, D. L., Asper, D. L., & Perry, H. M. (2003). Ecological and economic implications of a tropical jellyfish invader in the Gulf of Mexico. *Biological Invasions, 5*, 53–69.

Gramitto, M. E., Lucchetti, A., & Colarossi, M. (2010). *Guida pratica all' attività di ispezione e controllo nella pesca marittima*. CNR-ISMAR Technical Report, Ancona, 178pp.

Grati, F., Scarcella, G., Polidori, P., Domenichetti, F., Bolognini, L., Gramolini, R., Vasapollo, C., Giovanardi, O., Raicevich, S., Celic, I., Vrgoc, N., Isajlovic, I., Jenic, A., Marceta, B., & Fabi, G. (2013). Multi-annual investigation of the spatial distributions of juvenile and adult sole (*Solea solea L.*) in the Adriatic Sea (northern Mediterranean). *Journal of Sea Research, 89*, 12–22.

Grove, M., & Breitburg, D. L. (2005). Growth and reproduction of gelatinous zooplankton exposed to low dissolved oxygen. *Marine Ecology Progress Series, 301*, 185–198.

Hay, S. (2006). Marine ecology: Gelatinous bells may ring change in marine ecosystems. *Current Biology, 16*(17), R679–R682.

Heckmann, R. (2004). What else can happen? Other problems for fish production. *Aquaculture Magazine*, Mar/Apr, 1–8.

Hsieh, Y.-H. P., Leong, F.-M., & Rudloe, J. (2001). Jellyfish as food. *Hydrobiologia, 451*, 11–17.

IREPA. (2012). *Osservatorio economico sulle strutture produttive della pesca marittima in Italia 2011*. Napoli: Edizioni Scientifiche Italiane, 252pp. http://www.irepa.org/attachments/article/320/Osservatorio%20Nazionale%202011.pdf

Jackson, J. B. C., Kirby, M. X., Berger, W. H., Bjorndal, K. A., Botsford, L. W., Bourque, B. J., Bradbury, R. H., Cooke, R., Erlandson, J., Estes, J. A., Hughes, T. P., Kidwell, S., Lange, C. B., Lenihan, H. S., Pandolfi, J. M., Peterson, C. H., Steneck, R. S., Tegner, M. J., & Warner, R. R. (2001). Historical overfishing and the recent collapse of coastal ecosystems. *Science, 293*, 629–637.

Kawahara, M., Uye, S., Ohtsu, K., & Iizumi, H. (2006). Unusual population explosion of the giant jellyfish *Nemopilema nomurai* (Scyphozoa Rhizostomeae) in East Asian waters. *Marine Ecology Progress Series, 307*, 161–173.

Kitamura, M., & Omori, M. (2010). Synopsis of edible jellyfishes collected from Southeast Asia, with notes on jellyfish fisheries. *Plankton and Benthos Research, 5*(3), 106–118.

Knowler, D. (2005). Reassessing the costs of biological invasion: *Mnemiopsis leidyi* in the Black Sea. *Ecological Economics, 52*(2), 187–199.

Kogovšek, T., Bogunović, B., & Malej, A. (2010). Recurrence of bloom-forming scyphomedusae: Wavelet analysis of a 200-year time series. *Hydrobiologia, 645*, 81–96.

Licandro, P., Conway, D. V. P., Daly Yahia, M. N., Fernandez De Puelles, M. L., Gasparini, S., Hecq, J. H., Tranter, P., & Kirby, R. R. (2010). A blooming jellyfish in the northeast Atlantic and Mediterranean. *Biology Letters, 6*(5), 688–691.

Lilley, M. K. S., Houghton, J. D. R., & Hays, G. C. (2009). Distribution, extent of inter-annual variability and diet of the bloom-forming jellyfish *Rhizostoma* in European waters. *Journal of the Marine Biological Association of the United Kingdom, 89*(1), 39–48.

Lynam, C. P., Heath, M. R., Hay, S. J., & Brierley, A. S. (2005). Evidence for impacts by jellyfish on North Sea herring recruitment. *Marine Ecology Progress Series, 298*, 157–167.

Lynam, C. P., Gibbons, M. J., Axelsen, B. E., Sparks, C. A., Coetzee, J., Heywood, B. G., & Brierley, A. S. (2006). Jellyfish overtake fish in a heavily fished ecosystem. *Current Biology, 16*(13), R492–R493.

Lynam, C. P., Lilley, M. K. S., Bastian, T., Doyle, T. K., Beggs, S. E., & Hays, G. C. (2011). Have jellyfish in the Irish Sea benefited from climate change and overfishing? *Global Change Biology, 17*, 767–782.

Malej, A., & Malej, A., Jr. (2004). Invasion of the jellyfish *Pelagia noctiluca* in the Northern Adriatic: A non-success story. Chapter 16. In H. Dumont, T. A. Shiganova, & U. Niermann (Eds.), *Aquatic invasions in the Black, Caspian and Mediterranean Seas* (pp. 273–285). Dordrecht: Kluwer.

Mariottini, G. L., & Pane, L. (2010). Mediterranean jellyfish venoms: A review on scyphomedusae. *Marine Drugs, 8*(4), 1122–1152.
Mills, C. E. (2001). Jellyfish blooms: Are populations increasing globally in response to changing ocean conditions? *Hydrobiologia, 451*(1–3), 55–68.
Mozetic, P., Solidoro, C., Cossarini, G., et al. (2010). Recent trends towards oligotrophication of the northern Adriatic: Evidence from chlorophyll a time series. *Estuaries and Coasts, 33*, 362–375.
Mulazzani, L., Camanzi, L., & Malorgio, G. (2012). Price formation and geographic market integration: An empirical investigation of Adriatic small pelagic species. *Fisheries Research, 119–120*, 99–107.
Nagata, R. M., Haddad, M. A., & Nogueira, M. J. R. (2009). The nuisance of medusae (Cnidaria, Medusozoa) to shrimp trawls in central part of southern Brazilian Bight from the perspective of artisanal fishermen. *Pan-American Journal of Aquatic Sciences, 4*(3), 312–325.
Natural England. (2010). *Monitor of engagement with the natural environment: The national survey on people and the natural environment* (Tech. Rep. NECR050). Sheffield: Natural England.
Nishikawa, J., Thi Thu, N., Ha, T. M., & The Thu, P. (2008). Jellyfish fisheries in Northern Vietnam. *Plankton and Benthos Research, 3*(4), 227–234.
Omori, M., & Nakano, E. (2001). Jellyfish fisheries in southeast Asia. *Hydrobiologia, 451*, 19–26.
Palmieri, M. G., Barausse, A., Luisetti, T., & Turner, R. K. (2014). Jellyfish blooms in the Northern Adriatic Sea: Fishermen's perceptions and economic impacts on fisheries. *Fisheries Research, 155*, 51–58.
Pauly, D., Christensen, V., Dalsgaard, J., Froese, R., & Torres, F. (1998). Fishing down marine food webs. *Science, 279*, 860–863.
Perrings, C. (2001). The economics of biological invasions. *Land Use and Water Resources Research, 1*(3), 1–9.
Purcell, J. E. (2005). Climate effects on formation of jellyfish and ctenophore blooms: A review. *Journal of the Marine Biological Association of the United Kingdom, 85*(3), 461–476.
Purcell, J. E. (2012). Jellyfish and ctenophore blooms coincide with human proliferations and environmental perturbations. *Annual Review of Marine Science, 4*, 209–235.
Purcell, J. E., & Arai, M. N. (2001). Interactions of pelagic cnidarians and ctenophores with fishes: A review. *Hydrobiologia, 451*, 27–44.
Purcell, J. E., Breitburg, D. L., Decker, M. B., Graham, W. M., Youngbluth, M. J., & Raskoff, K. A. (2001). Pelagic cnidarians and ctenophores in low dissolved oxygen environments: A review. In N. N. Rabalais & R. E. Turner (Eds.), *Coastal hypoxia: Consequences for living resources and ecosystems* (Coastal and estuarine studies, Vol. 58, pp. 77–100). Washington, DC: American Geophysical Union.
Purcell, J. E., Uye, S., & Lo, W.-T. (2007). Anthropogenic causes of jellyfish blooms and their direct consequences for humans: A review. *Marine Ecology Progress Series, 350*, 153–174.
Quiñones, J., Monroy, A., Acha, E. M., & Mianzan, H. (2013). Jellyfish bycatch diminishes profit in an anchovy fishery off Peru. *Fisheries Research, 139*, 47–50.
Richardson, A., Bakun, J. A., Hays, G. C., & Gibbons, M. J. (2009). The jellyfish joyride: Causes, consequences and management responses to a more gelatinous future. *Trends in Ecology & Evolution, 24*(6), 312–322.
Riisgård, H. U., Madsen, C. V., Barth-Jensen, C., & Purcell, J. E. (2012). Population dynamics and zooplankton-predation impact of the indigenous scyphozoan *Aurelia aurita* and the invasive ctenophore *Mnemiopsis leidyi* in Limfjorden (Denmark). *Aquatic Invasions, 7*(2), 147–162.
Scheffer, M., Carpenter, S., Foley, J. A., Folke, C., & Walker, B. (2001). Catastrophic shifts in ecosystems. *Nature, 413*(6856), 591–596.
Schiariti, A., Kawahara, M., Uye, S., & Mianzan, H. (2008). Life cycle of the jellyfish *Lychnorhiza lucerna* (Scyphozoa: Rhizostomeae). *Marine Biology, 156*(1), 1–12.
Sen, A., Harwood, A. R., Bateman, I. J., Munday, P., Crowe, A., Brander, L., Raychaudhuri, J., Lovett, A., Provins, A., & Foden, J. (2011). Economic assessment of the recreational value of

ecosystems in Great Britain. In: *UK national ecosystem assessment technical report.* Cambridge: UK National Ecosystem Assessment, UNEP-WCMC.

Sen, A., Harwood, A. R., Bateman, I. J., Munday, P., Crowe, A., Brander, L., Raychaudhuri, J., Lovett, A., Provins, A., & Foden, J. (2014). Economic assessment of the recreational value of ecosystem: Methodological development and national and local application. *Environmental and Resource Economics, 57*, 233–249.

Shiganova, T., Mirzoyan, Z., Studenikina, E., Volovik, S., Siokou-Frangou, I., Zervoudaki, S., Christou, E., Skirta, A., & Dumont, H. (2001). Population development of the invader ctenophore *Mnemiopsis leidyi* in the Black Sea and in other seas of the Mediterranean basin. *Marine Biology, 139*, 431–445.

Stohlgren, T. J., & Schnase, J. L. (2006). Risk analysis for biological hazards: What we need to know about invasive species. *Risk Analysis, 26*(1), 163–173.

Sugahara, T., Ueno, M., Goto, Y., Shiraishi, R., Doi, M., Akiuama, K., & Yamauchi, S. (2006). Immunostimulation effect of jellyfish collagen. *Bioscience, Biotechnology, and Biochemestry, 70*, 2131–2137.

UK National Ecosystem Assessment. (2011). *The UK national ecosystem assessment: Synthesis of the key findings*. Cambridge: UNEP-WCMC.

Uye, S. (2008). Blooms of the giant jellyfish *Nemopilema nomurai*: A threat to the fisheries sustainability of the East Asian Marginal Seas. *Plankton Benthos Research, 3*, 125–131.

Uye, S. (2010). Human forcing of the copepod-fish-jellyfish triangular trophic relationship. *Hydrobiologia, 666*, 71–83.

Van Ginderdeuren, K., Hostens, K., Hoffman, S., Vansteenbrugge, L., Soenen, K., De Blauwe, H., Robbens, J., & Vincx, M. (2012). Distribution of the invasive ctenophore *Mnemiopsis leidyi* in the Belgian part of the North Sea. *Aquatic Invasions, 7*(2), 163–169.

The manufacturer's authorised representative in the EU is Springer Nature Customer Service Centre GmbH, Europaplatz 3, 69115 Heidelberg, Germany. If you have any concerns regarding our products, please contact ProductSafety@springernature.com

Printed and bound by CPI Group (UK) Ltd, Croydon, CR0 4YY
30/09/2026
02225572-0007